THE HEALTH OF NATIONS

THE HEALTH OF NATIONS

The Campaign to End Polio and Eradicate Epidemic Diseases

KAREN BARTLETT

ONEWORLD

A Oneworld Book

First published by Oneworld Publications, 2017

ISBN 978-1-78607-068-5
eISBN 978-1-78607-069-2

Illustration credits
Introduction Opener: David Stowell/Geograph.org.uk. Chapter 1 Opener: Centers for
Disease Control and Prevention's Public Health Image Library. Chapter 2 Opener: US Food
and Drug Administration. Chapter 3 Opener: FDR Presidential Library & Museum. Chapter
4 Opener: Bill & Melinda Gates Foundation. Chapter 5 Opener: Centers for Disease Control
and Prevention/James Gathany. Chapter 6 Opener: John Oxley Library, State Library of
Queensland. Chapter 7 Opener: The Historical Medical Library of the College of Physicians
of Philadelphia. Chapter 8 Opener: John Moore/Getty Images. Chapter 9 Opener: World
Health Organization/PATH global health/Flickr.

Typeset by Falcon Oast Graphic Art Ltd.
Printed and bound in Great Britain by Clays Ltd, St Ives plc

Oneworld Publications
10 Bloomsbury Street
London WC1B 3SR
England

For my mother

CONTENTS

Contents

The grumpy country doctor, Edward Jenner, changed the course of medicine by vaccinating local villagers in his 'Temple of Vaccinia', situated in his idyllic English country garden.

INTRODUCTION

A Gloucestershire garden

Tucked away in the farthest corner of a country garden in Gloucestershire sits an unprepossessing summerhouse, shaded and quiet, calmly anchored to the cool English earth. There is scarcely a whisper that, in this spot, scientific innovation shook the world and changed the course of human history. This is the 'Temple of Vaccinia', a tiny mushroom-shaped building, with bark-covered walls and a thatched roof. Although its silence now speaks of quiet neglect, this was once a hive of activity, with a line of people snaking around the garden, waiting to be vaccinated against smallpox by their local doctor, Edward Jenner.

Edward Jenner was not the first person to understand that exposure to cowpox led to immunity against the far more serious disease of smallpox but he was the first truly to understand the consequences of vaccination and to seek to prove them and publicize it in the medical community. Much to his chagrin, he believed he received insufficient thanks or appreciation for his work in his lifetime. Astonishingly, his reputation derived more from his work in understanding how cuckoo chicks pushed other birds from the nest than from being the founding father of vaccination. Nonetheless, his discovery would prove to be a profound turning point; the beginning of a centuries-long endeavour to save millions of people from smallpox and other diseases. The success of vaccination gave rise to the idea that certain diseases could not only be controlled but wiped out entirely, as the eradication of smallpox, in 1979, would eventually prove.

The success of the smallpox campaign inspired legions of doctors, scientists and health workers to believe that disease *eradication* was possible. However, thirty-seven years later, smallpox remains the only disease to be successfully eradicated. It has taken decades to bring the world tantalizingly close to the eradication of another fearsome disease: polio, which once killed, crippled and ruined the lives of millions and struck terror into the hearts of families around the world. If polio is finally wiped out, the eradication of other well-known diseases, such as measles, could be within reach. Even the eradication of malaria is on the table.

The words 'polio eradication' sound clinical and medical, perhaps even underwhelming and technical. The truth, however, is costly, messy and sometimes heartbreaking. Success would be a staggering victory; polio eradication could inspire and embolden scientists to make other, even more momentous, breakthroughs, making medicine as inspiring for the twenty-first century as the moon landings were for the twentieth.

A world without just one or two of its terrifying epidemic diseases would be radically different to the one we live in. First world countries could save billions of dollars in health spending and foreign aid, while developing nations could see a boom in their productivity and economic output. Our already-bulging planet could have many more human beings to feed, house and employ.

The balance of global power would alter in ways we cannot yet envisage.

These arguments are not new and nor is the human effort to eradicate disease. This book explores the complicated history of vaccination and our efforts to tackle some of the diseases we have most feared. It tells the story through one disease, polio, and describes first-hand the final stages of one of the biggest health campaigns in history, as well as the stories of those who risk their lives to achieve what can seem an impossible aim.

Giving all children the polio vaccine has required creative solutions and massive collaboration. To reach remote communities, the Global Polio Eradication Initiative (GPEI) has delivered vaccine to some of the most inaccessible parts of the world, using helicopters, motorbikes, boats and camels. The scale of the task is mind-blowing: when polio still naturally occurred throughout India, each round of national immunizations involved 640,000 vaccination booths, 2.3 million vaccinators, 200 million doses of vaccine, 6.3 million ice packs, visits to 191 million homes, and the immunization of 172 million children. Innovations developed in India included house-to-house vaccine delivery plans and finger-marking. Messages were painted on the chimneys of brick kilns to encourage migrant workers to vaccinate their children; mobile health teams vaccinated nomadic populations; vaccination posts were established at borders and transit points, and in bus and railway stations; strategies that were used again on the Afghan-Pakistan border and in the Horn of Africa.

The polio eradication programme has worked in times of peace and war, developing strategies to vaccinate children during conflicts and humanitarian emergencies and establishing 'Days of Tranquillity' to interrupt conflicts in the Americas and later in the Democratic Republic of Congo (DRC) and Afghanistan. In 2013, vaccine manufacturers delivered an astonishing 1.7 billion doses of the oral polio vaccine (OPV). Experts estimate the cost of the final push towards eradication between now and 2019 will be $7 billion.

Yet although polio has now been beaten back and is endemic in only three countries in the world, eradication is far from certain. Immense logistical difficulties, political intrigue, war and cultural and religious beliefs form immense obstacles to disease eradication; in the case of polio, this final goal has been set back decades.

The history of disease eradication campaigns demonstrates that the final stages are by far the hardest. An almost Herculean effort of will is required to maintain the immense momentum and funding needed to be channelled towards a tiny number of cases. The history of disease eradication in the twentieth century is largely the story of falling at the final hurdle. After billions of dollars, decades of work and the deaths of hundreds of health workers – deliberately targeted for their work in saving children from disability and death – the future of a world without epidemic diseases might come down to a team of polio workers determined to deliver the final few drops.

The medical missionary Bill Foege (centre) was crucial in developing the surveillance and containment strategy that played a key part in ending smallpox. Here, he celebrated the eradication of the disease in 1980, with two other former directors of the programme at CDC in Atlanta, Dr J. Donald Millar (left) and Dr J. Michael Lane (right).

1

THE HIPPIES WHO BEAT SMALLPOX

Smallpox eradication really is one of the greatest accomplishments in health in the twentieth century and those of us who have come after live with the most important legacy of smallpox eradication, which is infinite benefits from there being no more smallpox. All the people that would have got smallpox and would have suffered and all of the resources that would have been spent on managing smallpox, are now saved. So the great benefit of eradication is absolute prevention. It guarantees that you've got that stream of benefits, in perpetuity, into the future.

Tim Evans[1]

On Wednesday, 18 November 1863, as Abraham Lincoln was travelling by train up the eastern seaboard of the United States, he told his personal secretary, John Hay, that he felt weak and unwell. Those symptoms were followed by a high fever, a headache and then backache and within a week his skin had erupted in scarlet blisters. He was almost certainly suffering from smallpox. Had his symptoms worsened a couple of days earlier the course of American history might have been very different, for Lincoln was travelling to Pennsylvania, to deliver what became known as the Gettysburg Address: his pledge to save and preserve the United

States. Although Lincoln recovered, smallpox continued to wreak havoc across the globe for another century, claiming lives and terrifying people.

It would be another hundred years before the author of *One Flew Over the Cuckoo's Nest*, Ken Kesey, typed an editorial for his local newspaper in Eugene, Oregon that urged local people to gather at his farm, join the Peace Corps and get on a bus to go to West Africa to fight smallpox. Kesey might have been more famous for his cross-country, LSD-fuelled bus trip with the Merry Pranksters but the hippies who responded to his editorial were about to set off on an even more extraordinary adventure. As one remembered, it was about 'life, and living it for real, man'.

In 1963, the World Health Organization (WHO) started a smallpox eradication and education programme. Run from an under-funded office in Geneva and driven by the determination of one man, the American doctor and epidemiologist D. A. Henderson, the smallpox programme motored on little more than the fumes of 1960s' idealism. As Larry Brilliant, a young doctor who joined the campaign on the advice of his Indian guru, Neem Karoli Baba, remembered: 'We didn't have cell phones or phones; we barely had a photocopy machine. There were no computers.'[2]

Brilliant travelled to Asia on a bus bought from the proceeds from starring in a film, *Medicine Ball Caravan*. His bus eventually took him all the way from his ashram to the UN office. 'I had never seen a case of smallpox,' Brilliant told an interviewer in 2000:

> I had hair down the middle of my back and I was wearing a white robe. Everybody in the United Nations was over fifty and wearing a business suit ... I walked in and said: 'My mystic sent me to cure smallpox.' I was told to go

home. I took the seventeen-hour bus ride back
to the ashram and told Baba that I had failed.

After making the journey many more times, 'slowly, the robe gave
way to pants, then to a shirt, then to a tie, then to a haircut and then
to a CV,' Brilliant said. 'I learned to look like a diplomat.'[3] Not
everyone on the smallpox programme was as eccentric as Larry
Brilliant but many had something of a maverick spirit that drove
them on in the face of opposition, both on the ground and from
international agencies and institutions who believed they were
aiming for the impossible. It would take fourteen years but eventually, on 9 December 1979, the disease that had been a major killer
for centuries was wiped out through what one volunteer described
as 'common sense, teamwork and love'. Actually, it took a little
more than that.

Smallpox had long been one of the deadliest diseases in human history. When the mummified body of Pharaoh Ramses V, who died
in 1157 BCE, was discovered, his face was clearly ravaged by the
pockmarks of smallpox's horrific rash of pustules. By 1775 CE, it
was estimated that ninety-five per cent of the world's population
had been exposed to smallpox and one in seven had died from it.
For those who lived, the personal consequences were dire, including scarring and disfigurement, or 'rotting face' as the Native
Americans called it.

The political repercussions of the disease that Thomas
Macaulay called the 'most terrible of the ministers of death' were
equally profound. A smallpox infection brought to Central America
by an African slave killed millions of Aztecs, and in part accounted
for the successful Spanish conquest of Mexico. Smallpox wreaked
similar devastation in other countries, wiping out nearly forty per

cent of the population of Iceland in 1241, when the infection broke out on a visiting Danish ship. Smallpox claimed the life of Louis XV in France and ended the Stuart dynasty.

Once a person caught smallpox the progress of the disease was well-understood. Approximately ten days after infection, the sufferer began to feel very ill, with headache, fever and backache. Two days later, the fever subsided and a rash appeared, starting in the mouth, throat and face. For the next two weeks, the fever would return and the rash would spread across the upper body and down to the hands and feet. What began as flat spots turned into raised hard spots and eventually into soft pus-filled scabs. After two weeks, the fever receded and the scabs dried and fell off, usually leaving unmistakable pockmarks. If this was followed by a bacterial infection of the pockmarks, or an infection of the bone, the sufferer died.

If the progress of the disease was evident and devastating, understanding remained hazy. The eminent Baghdadi physician, Al Rhazi, argued in 910 that smallpox was due to a latent contagion in the body but was less serious than measles (he probably encountered the less serious strain of the disease). Some doctors and scientists, including the Dutch physician, Hermann Boerhaave, understood as early as the 1700s that smallpox was a contagious disease, spread from person-to-person, but this theory faced opposition from proponents of the 'miasma theory', who believed that the disease was one of many conditions caused by the 'clashing elements' of bad and foul air and could be cured by greater sanitation.

Not until 1931, when the invention of the electron microscope made it possible to see such tiny organisms, was the smallpox (Variola) virus described. Shaped like an 'undistinguished rounded brick',[4] it belongs to the same pox virus family as cowpox, vaccinia, monkeypox and taterapox (smallpox's nearest genetic ancestor, which exists only in wild gerbils). Unlike other pox viruses, Variola (in both its two strains, major and minor) affected only humans. This was an essential factor for an eradication campaign, as the disease could be wiped out in human hosts without the risk of it living on within an animal reservoir.

The concept of inoculation had been known in some parts of the world for centuries. India and Africa had their own customs to offer protection, while the Chinese understood that infecting children with a weak, mild form of the virus – using a silver tube to blow a dried sample up the nose – would inoculate them against catching a more serious and deadly form later in their lives. Famously, Lady Mary Wortley Montagu popularized the practice and brought it to European attention.

Born Mary Pierrepont in London, probably in early 1689, Lady Mary contracted smallpox in 1715, when she was twenty-six years old. The infection left her deeply scarred and stripped of her eye-lashes. Her disfigurement did not prevent her from scandalizing London society by eloping with Edward Wortley Montagu and joining him in Constantinople, where he was British ambassador. From that posting, Lady Mary wrote to her friend Sarah Chiswell that she had heard of a practice carried out on the women of the Sultan's harem in Circassia in the North Caucasus, which involved 'engrafting' the pox into the women, leaving the disease 'entirely harmless'.

> An article in the *Philosophical Transactions of the Royal Society* in 1714 describes the process of variolation, in which a surgeon scratched a patient's arm with a lancet until he drew blood, mixed those blood droplets with some

smallpox pus and then reapplied the mixture
to a fresh cut on the arm, binding it tightly to
make sure no liquid leaked out.

Lady Mary instructed Charles Maitland, the Scottish surgeon to the British Embassy, to 'variolate' her six-year-old son Edward. When that proved successful, he went on to inoculate her three-year-old daughter in London in 1721. This was the first known variolation carried out by a medical professional in England; it aroused the interest of the intelligent and well-read Princess of Wales, Caroline of Ansbach, who asked the king to order what became known as the 'Royal Experiment'.

In August that year six prisoners held under sentence of death in Newgate prison were offered a life pardon if they agreed to be inoculated. After a brief illness the condemned men, who had no doubt jumped at the chance of a reprieve, began to feel better, prompting the Princess of Wales to order a further experiment on six orphans of the parish in London. Eventually, she was satisfied that the practice would be safe for her own children. The procedure was successfully carried out on the eleven-year-old Princess Amelia, nine-year-old Princess Caroline and the teenaged Prince Frederick. Variolation became a popular practice in Europe, even though the fatality rate could be as high as one in five.

Despite the high death rate, variolation was seen as a vast improvement on any previous medical procedure connected with smallpox. Voltaire himself lauded Princess Caroline's efforts, writing that: 'This princess was born to encourage the arts and the well-being of mankind; even on the throne she is a benevolent philosopher; and she has never lost an opportunity to learn or to manifest her generosity.' (Voltaire's Eleventh Letter: On Smallpox Inoculation.)

The success of variolation would eventually grow dim, however, in the face of the safety and success of vaccination. The key difference between the two procedures is that variolation involved infecting a patient with a mild form of the human smallpox virus, which could potentially turn into full-blown smallpox or infect other people, but vaccination – as pioneered by Edward Jenner – used a sample of cowpox and so could not cause smallpox.

A country doctor

Like many of the other key figures in the history of disease eradication, Jenner was something of an oddball and outcast. He was a complex man whose many, and very varied, interests, combined with a difficult personality and his estrangement from the London medical establishment, meant that his truly great achievements were often overlooked, much to his immense frustration. Even today, the condition of his large but slightly gloomy and neglected Gloucestershire home gives the impression that the status of his work has yet to be fully appreciated.

Born in 1749 in Berkeley, Gloucestershire, Jenner was orphaned at the age of five and brought up by his sisters. His father had been the vicar of the small town and the family enjoyed the patronage of the influential Earl of Berkeley. They were neither the richest nor the most important family in the town; that privilege belonged to the Berkeleys, but life was very comfortable. After five years at boarding school, Jenner was apprenticed at the age of thirteen to a surgeon, Daniel Ludlow, in nearby Chipping Sodbury and spent six years learning about the life of a local doctor. Ludlow had himself trained at St George's Hospital in London and in 1770, his young pupil moved to the capital and spent two years living and studying with the eminent Scottish surgeon, John Hunter, who, with his brother William, was instrumental in teaching human anatomy using fresh human specimens rather than stylized drawings.

Although he could have stayed in London, in a pattern that would become familiar, after two years Jenner decided to return to Gloucestershire to begin life as a country doctor. In Berkeley, Jenner was quite the dilettante: writing poetry, dressing in the latest fashions and playing the flute. Life before his 'great discovery' was varied, fun and challenging and his scattergun achievements were not underrated (not least by Jenner himself, who always had a very high opinion of his talents). To satisfy his academic curiosity he joined the Convivio Medical Society, which met once a week at the Ship Inn in the nearby village of Alveston. He pursued various intellectual interests, including organizing one of the first unmanned flights by a hydrogen-filled balloon and producing a seminal study on how cuckoos infiltrated other birds' nests and pushed out the existing eggs. This discovery led to him being elected as a Fellow of the Royal Society in 1788, at the age of forty. Jenner deserves additional credit for being the first doctor to identify that angina was caused by narrowing and hardening of the coronary arteries. There is no doubt that he could have furthered his career and made more money by staying in London, but Jenner liked being a country doctor, which would ultimately be fortunate for humankind.

Jenner was familiar with variolation and indeed variolated his son Edward, two servants and numerous ordinary people in his care for free, when a smallpox epidemic threatened in 1789. Jenner's biographer John Barron claimed that a milkmaid had told the young Jenner: 'I cannot take that disease [smallpox] as I have had the Cow Pox.' While variolating milkmaids, Jenner recognized that this folk wisdom was true; when milkmaids were variolated, the result was often only a small blister, rather than the expected illness.

Jenner was not the first to understand the link between cowpox and smallpox. He had first heard of the connection decades earlier and made detailed sketches of cowpox pustules to discuss with colleagues in London in 1788. His colleague John Fewster wrote a paper for the London Medical Society in 1765 entitled 'Cowpox and its ability to prevent smallpox', which they discussed at the Convivio Medical Society. Neither was Jenner the first to understand that inoculating patients with cowpox would protect them from contracting smallpox; a Dorset farmer, Benjamin Jesty, made that connection in 1774, infecting two of his children and his wife with cowpox using a darning needle.

Jenner's contribution was to establish the connection and the procedure as a proven medical practice. Famously, he experimented on James Phipps, the eight-year-old son of his gardener, using cowpox pus from the hand of a milkmaid, Sarah Nelmes and from Blossom the 'Gloster' cow. This historic event took place on 14 May 1796 but Jenner took his time to come to his conclusions. This tardiness was another reason why his discoveries were often not given the attention they deserved.

Fortunately, the experiment was successful. James Phipps complained about discomfort in his armpit and felt generally unwell. Gradually, however, these mild symptoms faded and he made a full recovery. Seven weeks after inoculating Phipps with cowpox Jenner variolated him with smallpox. James had only a very small reaction, typical of the reactions of patients who were already protected against the disease. On 19 July 1796 Jenner wrote to his friend Edward Gardener: 'I have at length accomplished what I have so been long been waiting for, the passing of the vaccine Virus from one human being to another by the ordinary mode of inoculation.'[5] (Although Jenner uses the words 'vaccine' and 'virus' here, he did not understand either of them in the sense that we do today.)

Despite decades of prevarication, Jenner's discovery was an immense step. He was not correct about everything; for example he insisted that a single vaccine would last for a lifetime, although evidence emerged that it wore off as the years passed. However, his vaccine provided many years' immunity in a single dose and it became the workhorse of the smallpox campaign, used by the WHO in the eradication campaign of the 1960s and 1970s.

Sitting in Jenner's garden on a warm summer afternoon in 2015, Gareth Williams, professor of medicine and author of *Angel of Death: The Story of Smallpox*, confirmed:

> Jenner was not the first person to think about vaccination or even the first person to experiment with it but Jenner was decisive in that he wrote it up, he made his discovery accessible to everybody, whether they were medical or not. He didn't try to grab his secret, he wanted to spread the word and he pushed it, through force of character, into mainstream medicine. It's probably an exaggeration to say he was the father of vaccination because other people did think of it but he's the one that really made it happen, so he's the one that deserves the credit for that.[6]

Jenner wrote up his single experiment and sent two versions of the manuscript to the Royal Society in March and April 1797. His chosen recipient, Sir Everard Home, remained unconvinced and suggested to Jenner that more studies were needed. Chastened, Jenner withdrew his report and repeated the experiment on several more subjects. The results were successful but Jenner decided to publish his report privately, rather than resubmitting it to the Royal Society. Much to his dismay, his work received little attention. Jenner might have despaired, but his friend Henry Cline seized on his discovery and publicized it until the London medical establishment began to take notice.

Jenner believed he was snubbed by the establishment and he faced many challenges to his work, including those of his rival, the wily George Pearson of St George's Hospital, London. Pearson acknowledged Jenner's work but threatened to eclipse him as a vaccine provider. Ultimately, however, Jenner's discovery made him a very rich man. He was awarded two payments by Parliament of sums that would, in today's money, make him a multi-millionaire. Awards from Parliament were not unknown; in 1773 the clockmaker, John Harrison, was awarded eight thousand pounds (nearly a million pounds today) for the chronometer that enabled the accurate measurement of longitude. After much debate, Edward Jenner was awarded ten thousand pounds (supplemented by a further twenty thousand pounds) for being the 'sole discoverer' of the 'greatest discovery ever made in medicine'.

Recognition for Jenner was a slow and complicated process; he faced much criticism and considerable disagreement within the medical community. His vaccine also provoked a side-effect that remains a fact of modern life: the rise of the 'anti-vaxxers'. To many, injecting people with animal pus seemed horrifying. One doctor claimed that the smallpox vaccine gave rise to 'ox-faced babies', while another worried that ladies would be chased through the fields by sex-crazed bulls.

Vivid as these scenarios were, Jenner's procedure was nevertheless a success. By the early years of the nineteenth century smallpox vaccination had spread around the world, through a remarkable chain of arm-to-arm transmission, starting with the compulsory vaccination of Napoleon's army in 1805. A Viennese doctor, Jean de Carro, initiated a human arm-to-arm chain that carried the vaccine across Greece and Turkey and as far as India, where three-year-old Ann Dusthall was the first person to be vaccinated on Indian soil in Bombay in 1802. Centuries before the concept of global healthcare emerged in institutions like the WHO, arm-to-arm transmission was a striking example of colonial reach. King Charles IV of Spain commissioned the royal physician, Francisco Xavier Balmis, to undertake a voyage carrying twenty-two vaccinated foundling children across the Spanish Empire, including the Canary Islands, South America and the Philippines and the Portuguese followed a similar model in Bahia in Brazil.

In North America, vaccination began in Canada in 1798, when Jenner's childhood friend John Clinch, who lived in Newfoundland, received a package of dried vaccine. Vaccination got off to a difficult start in America, when a Boston physician, Benjamin Waterhouse, conceived an ill-fated plan to make money from the vaccine, which backfired when poor-quality vaccines led to sixty smallpox deaths. Nevertheless, President Thomas Jefferson was convinced of the vaccine's merits after conducting experiments at his home, Monticello, in Virginia, and went on to spread vaccination throughout Virginia, Philadelphia and Washington DC. In 1804 Jefferson wrote to Edward Jenner:

> You have erased from the calendar of human
> afflictions one of its greatest. Yours is the com-
> fortable reflection that mankind can never forget
> that you have lived. Future nations will know
> by history only that the loathsome smallpox
> has existed and by you has been extirpated.[7]

Throughout the nineteenth century, and into the twentieth, smallpox vaccination continued to confer immense benefits in any location where it was successfully deployed: smallpox cases in the USA fell from 102,000 in 1921 to none by 1949, and in Europe cases fell from 122,000 in 1921 to 281 in 1950. The problem was that such successful locations were severely limited. In the rest of the world, the numbers of smallpox cases remained high. While Edward Jenner and Thomas Jefferson dreamt of a smallpox-free world, the reality was that world-wide vaccination was extremely difficult. Successfully carrying the vaccine across vast, difficult terrain, to every remote corner of the globe, and vaccinating every person, was practically impossible.

Some major obstacles were overcome. By the mid-nineteenth century, the innovation of using cows and heifers as an alternative to arm-to-arm injections made it easier to keep the vaccine chain going. Even so, there was little vaccination in sub-Saharan Africa, where an epidemic claimed 2,500 lives in Cape Town in 1841. In the twentieth century nearly half a million cases were reported during an epidemic in Indonesia in 1949, while 230,000 people were killed in an outbreak in India in 1944.

Often the challenges were human, not geographic or logistical, with some people unconvinced of the benefits of vaccination. In Nigeria the Yoruba people produced wooden statues of a smallpox god, while the goddess of smallpox, Sitala or Sitala Mala, was worshipped by Hindus, Buddhists and tribal people across India. Sitala – meaning the 'cool one' – scattered the seeds of smallpox but also brought relief through cooling water and a cleansing brush. Worship of Sitala remained a problem for vaccinators in India well into the final eradication campaign of the late twentieth century.

Even in Britain, smallpox vaccination brought significant challenges. Despite its high death rate, variolation remained so popular that the government decided it could only be ended by a ban, and introduced a law to prohibit variolation in 1840, which was complemented by free vaccination of the poor to be carried out by the Poor Law Commissioners. However, as the commissioners were highly unpopular, and usually poorly paid and badly trained, uptake of vaccination was slow. Compulsory vaccination, introduced in 1853, was highly unpopular and led to the establishment of the Anti-Vaccination League, a branch of which was also established in the USA.

Opposition was strongest in the city of Leicester, where four thousand people faced prosecution in 1885 for failing to vaccinate their children, and more than twenty thousand attended a protest meeting. Eventually Leicester developed its own method of smallpox containment, involving surveillance, isolation and good hygiene, and the vaccination of only the close contacts of those who had contracted the disease. (A comparable method was eventually used in the smallpox eradication campaign.) Sadly, a similar anti-vaccination campaign in Gloucester had much less positive results. With less emphasis on good hygiene surveillance and isolation, an epidemic that ripped through the city in 1895 killed more than four hundred people.

A post-war dream

Jenner's discovery faced strong opposition and for two hundred years vaccination remained sporadic across the world. Yet the importance of vaccination was beyond doubt and the concept was firmly established as a highly successful medical practice. The next leap forward in combatting the disease came in the mid-twentieth century, when doctors and public health officers began to revisit Jenner and Jefferson's dream: the complete eradication of the disease. Despite optimism and lofty idealism, it was undertaken with trepidation and uncertainty.

In 1950 Fred Soper, director of the Pan-American Sanitary Bureau (of whom more later) proposed the eradication of smallpox in the Americas. According to D. A. Henderson, who later directed the international smallpox eradication campaign, Soper had 'no interest in smallpox at all' but saw it as a means of engaging the US government with international health work and with the newly formed WHO. Smallpox was high on the American agenda, following an outbreak that terrified the nation when, in 1947, an American merchant, Eugene Le Bar, boarded a bus in Mexico City and carried the disease all the way to midtown Manhattan. Le Bar's diagnosis caused panic and turmoil and led to one of the biggest and most astounding public health campaigns: the vaccination of six million Americans, most within two weeks.

Soper believed that smallpox could be eradicated in the Americas using a new freeze-dried vaccine he had seen under development in Europe, and that American support for his effort would lead to increased support and budget for his other eradication efforts for malaria and yaws. Soper's belief in the freeze-dried vaccine was correct and a widespread campaign meant that by the mid-1960s smallpox was eradicated in the Americas, with the exception of Brazil. However, it was an intervention from an unexpected source, the Soviet Union, which first raised the possibility of global eradication.

After nine years' absence, the Soviet Union and its Communist allies rejoined the WHO in 1957. One of the WHO's first acts, at the Eleventh World Health Assembly in Minneapolis in May 1958, was to propose that it undertake a global smallpox programme. The Soviet deputy health minister, the virologist Victor Zhdanov, quoted Thomas Jefferson and called on the WHO to undertake mass vaccination campaigns in conjunction with the 'Leicester method' of surveillance and isolation.

The proposal was enthusiastically approved by other countries, keen to display solidarity. However, over the next seven years, progress was patchy. By 1966, the number of countries of the world with endemic smallpox was halved but the regions and countries that remained were the most difficult: West Africa, East Africa, India, Pakistan, Indonesia and Brazil. The WHO had limited funds and voluntary contributions by the countries themselves were small. Recognizing that the campaign needed new impetus, the 1966 World Health Assembly asked the director general of the WHO to draw up and present a ten-year smallpox eradication plan. The proposal had two essential components: systematic vaccination, and a surveillance and containment plan requiring weekly reports of cases from all health facilities and containment of outbreaks by special containment teams. The proposal called for $2.4 million of funding per year from WHO, and further voluntary contributions.

The idea itself, however, was divisive and politically charged. Many countries remained far from convinced. Some questioned whether eradication was even possible, while others were reluctant to commit the sums needed to achieve it. Four previous eradication efforts – hookworm, yellow fever, yaws and malaria – had failed. Malaria had been the subject of an intense campaign that cost more than $2.5 billion. Such failures led many in the global health community to shift their focus from disease eradication to the provision of basic healthcare.

Dr Donald Ainslie Henderson (always known as D.A.) told the WHO Bulletin how the vote proceeded and why he ended up with the seemingly unenviable task of running the programme:

> There was considerable debate as to whether
> [it] was a good idea or not. Several countries
> felt that it was impossible to do the job and

some were reluctant to provide more money to WHO to accomplish this. The Director-General, Marcolino Gomes Candau, was very much opposed, because the malaria eradication programme wasn't doing very well. His view was that if WHO were asked to undertake a second eradication programme, it would fail and that would reflect very badly on WHO and the public health community. He felt that the USA had played an important role in the debate in the assembly in persuading delegates to vote for this so he asked that an American – and specifically me – be assigned to the job, so when the eradication effort failed, the responsibility for it would be seen partly as that of the United States.

When the smallpox eradication plan was put to the vote, the World Health Assembly was split, deliberating for three days before finally deciding to support smallpox eradication by the narrowest margin: two votes. The global eradication campaign was born in uncertain circumstances.

A medical missionary

'With most global health programmes the science is actually a small part of it, it's the management that turns out to be the big part,' Bill Foege said:

We like to think we're good managers because we have military people that know about logistics. But we have somehow been afraid to put our best managerial abilities into global health, we don't think that way. And there's yet another lesson for smallpox that applies to polio and other diseases – which is you can't do any of these things unless you know the truth. And knowing the truth becomes such a significant thing: it needs measurement, it means knowing about things even when they hurt us. But we've also learned with both smallpox and polio that culture turns out to be a big area of science. And I learned early on in graduate school [that] if you tangle with culture, culture will always win.[8]

Bill Foege was born the son of a Lutheran minister but his thoughts turned from religion to medicine when, at fifteen, he was encased in a body cast for three months. With nothing to do but read and ponder the future, he began to think about working in Africa. That future would involve developing a very successful new strategy for disease control and eradication, honed by years of working on the ground.

After graduating from medical school, the six-foot-seven 'medical missionary' first worked in Epidemic Intelligence Services at the Centers for Diseases Control and Prevention (CDC) in Atlanta before leaving for a short posting with the US government-funded Peace Corps in India, where, for the first time, he encountered smallpox, and the notion that in most parts of the world people do not present themselves at a clinic with a single illness but with a variety of conditions that need untangling and treating. When he returned to the USA, Foege applied to the Harvard School of Public Health to study under the Nobel Laureate, Tom Weller, to consider how to use the training he had received to the benefit of the developing world.

'The die was being cast.' Foege said:

> Wolfgang Bulle, a surgeon for many years in
> a church-sponsored programme in India, had
> become a convert to public health. He agreed
> to sponsor a programme to emphasize public
> health and prevention in Nigeria. This would
> have been my lifetime's work, except that
> the Nigerian–Biafran civil war disrupted the
> programme and, while waiting for peace, I
> became obsessed with the possibilities of
> smallpox eradication.[9]

Working under the overall command of D. A. Henderson, then
heroically wrestling with the WHO bureaucracy in Geneva,
Foege's work became an essential part of the architecture for the
global smallpox eradication campaign. Although Africa was initially considered the most challenging continent for a smallpox
eradication campaign, Foege's work in East Nigeria established the
surveillance, reporting and isolation techniques that were not only
instrumental in beating smallpox but would become fundamental
for the polio eradication campaign.

The smallpox campaign in Nigeria was combined with a CDC
USAID measles programme, in which vaccination centres and
teams of mobile vaccinators moved from village to village, vaccinating children against measles and everyone against smallpox.
Local people viewed the campaign favourably, as they had received
a highly successful antibiotic treatment programme for yaws. Up
to one thousand people at a time lined up to pass through the vaccination stations, where they were vaccinated with a jet injector

that used high pressure air, rather than a hypodermic needle, to push the vaccine under the skin.

This mass vaccination campaign appeared to be working very well until, unexpectedly, an epidemic broke out in an area that was believed to have high vaccination levels. An investigation revealed that a particular religious sect had been avoiding vaccination and thus spreading smallpox throughout the region. Bill Foege became convinced that the mass vaccination strategy had failed and should be replaced by the surveillance and containment strategy he had developed in eastern Nigeria, tapping into a local network of missionary contacts who visited local villages by day and reported any smallpox symptoms by night. Instead of a mass 'blitz' of vaccinations, they would identify every village and case and work outwards, to target those who might be affected.

The benefits and drawbacks of a mass vaccination plan target-ing everyone, rather than a surveillance and containment strategy targeting a select few, would be hotly contested. Foege firmly believed in surveillance and containment and his first convert was the then regional director for eastern Nigeria, Dr A. Anezanwu. In truth Anezanwu needed little persuading; the strategy seemed sound and he was a member of the Igbo tribe, which lived in an area that was soon to declare independence from Nigeria and fight to establish Biafra as an independent country. Pursuing an indepen-dent strategy to Nigeria had political appeal.

The tension that would lead to war between Nigeria and Biafra was evident when the Nigerian government decided to halt vaccine supplies to the east, where Foege was working, until the rest of the country had caught up. Determined not to let this stand in his way, in March 1967 Bill Foege drove to Lagos to pick up jet-injector parts, and secretly filled the team's white pick-up truck with vaccines at the same time. It was a fortuitous move, as Foege was trapped in Accra in Ghana while war raged between Nigeria and Biafra; he was not able to return until 1968. When he was finally able to return, Foege was delighted to see that, despite two and half years of fighting, there were no reported cases of smallpox in eastern Nigeria. The surveillance-containment plan had worked.

Hard times

An effective vaccine, a global commitment, dedicated workers and a successful new surveillance and containment strategy meant that by the early 1970s smallpox was in decline almost everywhere. In early 1971, West and Central Africa reported their last case of small-pox, followed by Brazil nine months later and Indonesia in January 1972. Despite such progress, however, one region – India – made frustratingly slow progress.

The Indian authorities were more conservative than those in West Africa and more reluctant to adopt a surveillance-containment strategy, preferring mass vaccination campaigns. Mass vaccination had proved extremely successful in India, dramatically reducing numbers of smallpox but some areas were plagued by vaccines that did not 'take' properly, and so were not effective, while many rural areas remained inaccessible.

Greater use of freeze-dried vaccine and the development, in 1968 of a bifurcated needle, greatly reduced problems with the vaccination. The bifurcated needle, which was much easier to use, had two prongs with a drop of vaccine held between them. The vaccinator used the needle to make fifteen small scratches in the skin, efficiently and quickly offering protection to thousands of people. Yet despite such advances, it became apparent that smallpox was alarmingly under-reported, while official policy remained stubbornly resistant to abandoning the mass vaccination strategy.

For four years, from 1970, the Indian government increased smallpox funding fourfold, introducing mobile squads and establishing a ratio of one vaccinator for every 25,000 people. Yet despite these improvements, the number of cases of smallpox appeared to be going up. In 1971 the number of smallpox cases rose by twenty-five per cent and by another fifty per cent in 1972. However, it soon became clear that there were not more cases of smallpox, rather that better reporting methods of surveillance and containment were revealing the true extent of the problem.

By 1973, it was clear that the smallpox eradication campaign faced enormous challenges, particularly in the states of Uttar Pradesh (UP) and Bihar. Conditions were considered particularly difficult in Bihar, a rural state whose 56 million people still largely lived under tribal customs. (One Indian health worker told me most of her fellow countrymen considered Bihar so poor and backward they'd happily have lopped off the whole state and donated it to Pakistan.) Vast areas of Bihar were inaccessible at various times of the year and it had a large mobile population of workers whom no one had ever been able to adequately track. In the hot summer of 1973 the new smallpox strategy began in earnest, as vaccinators conducted six days of intense searching, blitzing villages and vaccinating the twenty nearest contacts of any smallpox case they discovered. Within a month six thousand new smallpox cases were discovered in UP and four thousand in Bihar. Smallpox was present in a shocking ninety per cent of districts. By March 1974 there were six hundred outbreaks a week. There were 67,000 new cases of smallpox in the first months of 1974; the campaign was reaching a tipping point and a stand-off with the Indian government was imminent. What came next was crucial.

Alarmed by the rising number of cases, the minister of health for Bihar believed that the surveillance-containment strategy had failed, leaving too many children unvaccinated. He argued that the only course of action was a return to the mass vaccination strategies of the past. In a meeting in Patna on 27 May the minister told public health workers that the battle was being lost and he intended to revert to mass vaccination immediately. The crowded room, filled with people who had dedicated themselves to the smallpox campaign for years, fell silent as they considered the possibility that their efforts might come to nothing. Then, a young local doctor raised his hand. In his rural village, he said, his neighbours would put out a fire by pouring water on the house that was ablaze, not wasting it on all the other houses in the village in case they caught fire too. The health minister considered this scenario and reluctantly conceded that the surveillance-containment strategy could have one more month to prove its success. All those working on the smallpox campaign breathed a sigh of relief.

A month was all that was needed. Workers had been clearing outbreaks at the rate of eight hundred a week and were working at full stretch. In May the Bihar figures revealed that outbreaks had fallen by one-third and the number of individual cases was also dropping dramatically. The tide was turning in UP too, where cases peaked at eight hundred new outbreaks a week in June 1974.

Within a year the blackspots of smallpox outbreaks that had hovered over the map of Bihar and UP had been wiped out. The Indian campaign had been intense, and at times faced fierce resistance. Local police had been deployed with the vaccinators for dawn raids on villages. Dedicated workers had travelled on camels, mules and bullock carts, crossing seemingly impassable rivers to vaccinate the most remote regions. But the finishing line was in sight. On 24 May 1975 a thirty-year-old homeless beggar, Saiban Bibi, fell ill in Assam. She was to be the last reported case of smallpox in India.

Bibi's recovery was a victory for the campaign in India but she represented another thorny problem; it appeared that she had contracted the disease from the refugee populations who moved back and forth across the border with Bangladesh. Smallpox recognized no borders. It was particularly vexing, as Bangladesh had rid itself of smallpox as early as August 1970, when it was still known as East Pakistan. Unfortunately, the 1971 war of independence from Pakistan, and the influx of large numbers of refugees, brought the disease back with a vengeance. The smallpox campaign was forced to begin again.

By the spring of 1974 the smallpox campaign in Bangladesh seemed to have regained lost ground but heavy rains brought devastating floods, ruined the rice crop and caused famine. Hundreds of thousands of refugees were on the move, spreading smallpox in their wake. One of the health workers flown in to help was a doctor from Bristol, Chris Burns-Cox. Burns-Cox, a former medical officer for the British colonial office in Borneo, joined the smallpox campaign in Jamalpur district on a short-term contract in July 1974:

> My job was to go round and check how well the village teams were working. In the place I'd been sent to they failed to report some cases, so things weren't working perfectly. They were in a mess. This was the monsoon and they were starving, wet and very, very poor. Morale was pretty low.[10]

Burns-Cox led a team of five other health workers who had originally been working on malaria eradication. They were 'pretty depressed,' he reported. 'It gets dark before seven o'clock and there was no point in staying up because there was no electricity in most of the villages. I remember lying there in the mud at night and the team leader saying: "Bangladesh is an empty bucket, there's nothing left".' According to Chris Burns-Cox, his colleagues felt that everything had gone wrong and Bangladesh was a dying nation: 'The monsoon was worse than usual, so a lot of people had lost their paddy crops and you could see the rice plants going black in the water. In every house I went into there was somebody waiting to die, saying "please let me die".'[11]

Burns-Cox and his team visited villages on market day, or when there was a celebration, holding up cards with photos to see if anyone recognized the symptoms of smallpox. 'If people said they had seen the symptoms we would go with that person to the house to see if it was smallpox, or whether it was chickenpox.' After recording his diagnosis, Burns-Cox and his team vaccinated as many people as they could find within a one-hundred-metre radius:

> Sometimes the women would run. If they knew I was coming they'd rush out of the houses and run into the jute plants and hide, so I had to wade in after them to catch them. I vaccinated people all over the place; on the train, up trees. At one site, there was a house that the team wouldn't go into. They told me the woman had just come out of prison and was dangerous. A man isn't supposed to go into a house if there's a woman in there without her husband but I was very cheeky, I just nipped in. The woman ran around in the middle of the house, trying to escape me. But eventually I got her in a half Nelson and vaccinated her. When I came out they said 'Do you know why she was in prison?' She'd just murdered her husband.[12]

Burns-Cox came to believe that the smallpox campaign could have only been driven and coordinated by US-led efforts. 'Only the Americans could have done it,' he said:

> When you were in the head office in Bangladesh you weren't allowed to say any-thing that wasn't to do with smallpox. You weren't allowed to discuss anything at all. It was smallpox, smallpox. The British would have said 'Oh, is there a Test match on – let's stop talking about smallpox and listen.'

In January 1975 the smallpox campaign in Bangladesh suffered a further blow when the government decided to demolish the ille-gal slums of Dhaka, turning half a million people into refugees. These people returned to their original homelands, carrying smallpox with them. But this was the final crisis; by February the Bangladeshi government had recognized the scale of the problem and authorized an emergency effort. Two hundred international experts from nineteen countries arrived to join forces with local health workers, determined to make a final push to eradicate small-pox in a country beset by the gravest problems. In November 1975 Bangladesh reported its last case of smallpox: Rahmina Banu, a three-year-old girl from Bhola Island, the largest island in the mouth of the Meghna River.

The last man to suffer from smallpox

With smallpox eradicated on the Indian subcontinent, it remained only in the eastern region of Africa and Ethiopia, where civil war had made the campaign particularly hazardous. Fighters had shot down an aid helicopter and some workers had found themselves under grenade attack. With resources freed from the campaign in India, the smallpox campaign made renewed efforts in Ethiopia and believed they were on the brink of announcing the last case of smallpox in the world, until news broke of an outbreak in Mogadishu. Investigations revealed that smallpox outbreaks in Somalia had been hushed up to save the government from embarrassment; the guilty party was removed from his post and a new campaign manager was introduced. Vaccination levels rose to ninety per cent and the disease was finally isolated to a remote area between the armed conflict in Somalia and Ethiopia. The nomadic people who lived there had shunned outsiders for years and were difficult to monitor and vaccinate but by 1977 the region had become the focus of a global campaign. Relentless focus on a smaller and smaller group of people paid off and smallpox dwindled to virtually nothing.

Finally, in October, the 'last case' appeared. A 23-year-old hospital cook in Mogadishu, Ali Maow Maalin, jumped into a jeep and offered to guide a health worker who was transporting a mother and two sick children to an isolation hospital. Ali had been vaccinated but the vaccination had not 'taken'. When he began to feel ill nine days later he was initially treated for malaria. Only after he was sent home from hospital did a visiting nurse recognize that his symptoms were those of smallpox. A frenzied vaccination campaign was organized for fifty thousand local people but no further cases were reported. Ali Maow Maalin was the last reported case of smallpox in the world. He recovered from the disease and used his experience to work for others, becoming a polio vaccinator. In the course of his work with the polio campaign in East Africa, Ali Maow Maalin contracted malaria in 2013, from which he later died, a human symbol of how diseases and disease eradication campaigns are fundamentally, and sometimes tragically, interlinked.

A message to the future

On 9 December 1979, a global commission made the historic pronouncement that smallpox was eradicated. The pronouncement was endorsed at a meeting of the World Health Assembly on 8 May 1980. Since then, the world has lived with the narrative that smallpox is the *only* disease to have been eradicated. Sceptics still claim that it is the only disease that will *ever* be eradicated. Undoubtedly, the costs of the campaign were huge and the effort intense.

As Larry Brilliant remembered:

> There were times when we almost lost with smallpox eradication. When there was an outbreak of 188,000 cases in India it was easy for the government to budget for smallpox programmes. When you are down to five cases the pressure to move that money to malaria, or something else, is immense.[13]

In the final stages of an eradication campaign, the cost is always high compared to the small number of remaining cases but as Bill Foege pointed out, the economic benefits of eradication live on:

> With smallpox eradication, the US investment is recouped every three months. What the US saves every year is actually greater than our dues for WHO. So it's not just human misery we're preventing, it's a financial asset. It's a different world.[14]

Foege saw smallpox eradication as a rare opportunity to invest in the world's children and grandchildren. As science improves, more of those opportunities will become available, whether we embrace them or not. In fact, many lessons of the smallpox campaign would prove directly applicable to the subsequent polio campaign, although polio would prove to be more complicated – requiring for example a global network of laboratories, to identify the disease.

Perhaps the most practical outcome of the smallpox campaign was the success of the detection and surveillance strategy developed by Bill Foege. The former director of the US National Immunization Program and member of the smallpox team, Walter Orenstein, said: 'It's not the numbers of kids immunized but the right kids immunized – and this strategy has played a big role in the polio campaign by using surveillance of the disease to detect problem areas and then take action.'[15]

But the end of smallpox was so much more than a practical victory: the disease that had killed and ravaged for centuries was gone and for those who championed eradication it was an unparalleled leap forward for global health. It spoke to the future: 'After we eradicated smallpox kids told their parents "I don't want to go to business school, I want to be a doctor because I want some of what they tasted when they eradicated smallpox",' Larry Brilliant said. He may no longer wear a long robe, or live in an Ashram but the spirit of smallpox eradication fuelled a sense of idealism for life.

> That's the same for polio. When we eradicate polio it will change the narrative that smallpox is the only disease that can be eradicated. That's crap. Smallpox is not the only disease that will be eradicated. If we fail to eradicate polio, we lose the opportunity to inspire the next generation. And when we eradicate polio, it will embolden a generation. So we can't fail.[16]

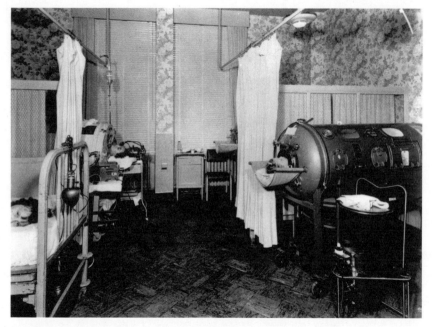

Paralytic polio struck terror into millions of parents, facing the grim
prospect of their infected children spending years confined in 'iron lungs'
to help them breathe.

2

THE CRIPPLER

We'll never end pandemics if we can't get rid of polio, there will be no public will for it.

Larry Brilliant[1]

Christine Wright was four years old when a polio epidemic broke out in south London. It was 1944 and the cold, bleak city was in its fifth year of war. Although her father was a policeman and her mother a commercial artist, Christine's family was poor and they lived with her grandmother in a flat in New Cross. 'I don't remember feeling any symptoms – not like I had the flu or anything,' she said. A local doctor advised her mother that she probably had childhood rheumatism and confined her to six weeks in bed. Within a few days, however, she started to lose the feeling in one leg. 'My mother made me run around the table, around and around. She didn't believe my leg was paralysed and thought I was pretending. I remember how hard it was to run around that table when one of my legs didn't work.'[2] Soon, an ambulance arrived and took the terrified little girl away, on her own, to an isolation hospital. Christine still remembers the excruciating pain as the doctors performed a lumbar puncture to determine whether she had polio. The result was positive. Christine was suffering from a disease the newsreels and radio broadcasts simply called 'The Crippler'.

For a few days Christine was confined in what she remembers as a 'glass cage', surrounded by other children in their own glass cages, before she was moved again to a recuperation hospital deep in the Surrey countryside. There, she lay in bed staring up at the high ceiling, where someone had painted rabbits for the disabled children to look at. Although she had hydrotherapy and a partial leg brace, the damage was done. Six months later, when Christine's parents came to collect her, they had to hold her up, one on each side, as they made their way down the drive of the hospital to wait for the bus back to London.

'My mother was very ashamed of me,' Christine said. 'Having a disabled child was a huge stigma.' Christine was sent to a special school for disabled children; her mother worried that the neighbours would see her getting on the bus. Eventually, the family moved to another part of London and Christine went to an ordinary school, where she worked hard, passing the exam to get into grammar school and eventually going to university. 'I always thought if I achieved these things my mother would be proud of me but all she ever said was "You having polio ruined my life."'

Despite such physical and emotional adversity, Christine graduated from university and became a teacher. She married, had two children and lived a full and happy life. For many years she walked unaided but soon after she reached sixty she noticed a marked deterioration in her condition. She visited her doctor, concerned that she had another disease, such as multiple sclerosis. In fact, she had 'post-polio syndrome', something which commonly affects older polio sufferers. This was a consequence of polio that Christine had not imagined and it seemed doubly cruel that people who had overcome disabilities earlier in their lives should be struck down again in their later years.

Christine can now walk only with the support of crutches. Yet despite the fact that she is struggling with the destructive after-effects of polio, Christine believes she is one of the lucky ones: 'I mostly feel very fortunate to have contracted polio in Britain where I could use my brain and go on and have a full life,' she said. 'I feel desperately sorry for people who catch it in other parts of the world, because once they have it their life is over.'

In much of the world stories such as Christine's, like polio itself, have been consigned to the history books. Thanks to the hugely successful vaccination campaign of the 1950s and 1960s most Americans and Europeans look puzzled, cast their minds back to a dim and distant memory and ask: 'Does polio still exist?' Yet until very recently the answer in the rest of the world is 'yes', with devastating consequences.

Currently, the disease is endemic in only three countries: Nigeria, Afghanistan and Pakistan: the end game is on. Memories may be fading but the importance of the polio eradication campaign extends well beyond the technocratic language sometimes employed by NGOs and healthcare organizations. The success or failure of the battle to wipe out polio completely will define twenty-first-century healthcare and determine the future of other, even more terrifying, viruses.

A remarkable team, combining scientists and tacticians at CDC, the World Health Organization, UNICEF and Rotary International with the philanthropy of the billionaire, Bill Gates, has wiped out endemic polio from almost all the world and almost entirely extinguished it from the collective memory of anyone living in Europe or the USA. If this team can succeed in eradicating the final one per cent of cases it will be remembered as having achieved a scientific breakthrough equivalent to putting a man on the moon, succeeded in only the second eradication of a disease in human history and emboldened future generations to make further breakthroughs in disease eradication. If they fail, critics will point to the thirty-year, US$8.6 billion campaign and claim it has spent a staggering amount of money and expended huge efforts in chasing an impossible dream; money and effort that could have been better spent on more achievable health efforts. When the world is gripped with worry about the spread of uncontrollable diseases, such as Ebola and other viruses, such a conclusion would be devastating. Yet war and insecurity make the goal of wiping out the tiny number of remaining cases of polio frustratingly elusive; time is ticking towards the campaign's self-imposed deadline of 2019 (now likely to be pushed back to 2020). In March 2014 the World Health Organization admitted the scale of the problem, announcing: 'The fight to eliminate polio is now imperilled . . . by insecurity, targeted attacks on health workers and/or a ban by local authorities on polio immunization.'

'We must eradicate polio,' said Larry Brilliant, who worked on the polio campaign after his years in India with the smallpox team and is now president of the Skoll Global Threats Fund. Despite the polio campaign's breathtaking technology and resources, Brilliant sees the same spirit of idealism and endeavour that characterized the efforts to eradicate smallpox forty years earlier: 'It's not just this one disease and all of its human pain and suffering that's at stake – it's anything we might do in the future as a global health community.'[3]

The polio eradication campaign has employed every element to fight the disease: the most advanced technology in the world to map communities and track volunteers combined with the oldest forms of transport, including canoes and camels, to transport vaccines and a large element of the human persuasion needed to convince politicians, religious leaders and parents to accept the countless rounds of oral doses needed to make the vaccine successful. It is also a campaign that many people are willing to risk their lives for. Polio vaccinators have been targeted and killed by rebel groups in Pakistan, Nigeria and the Horn of Africa. In 2013, more polio volunteers and workers were murdered than people died of the disease.

'In the smallpox programme in India we had 150,000 workers. They were the heroes,' Larry Brilliant said.

> In the polio programme in India, they had more than four million volunteers or workers. They were the heroes. And particularly in Pakistan, the vaccinators who risk their lives to go and do the simplest of things to protect the child from a crippling disease; they're the heroes.[4]

The lure of playing a part in what could be the biggest leap forward in global health in the twenty-first century seems irresistible, but the risks are very high. When a hundred so-called 'polio-busters' (nurses, doctors and health visitors) from around the world gathered at the CDC in Atlanta in the summer of 2014 for three weeks of intensive training before going into the field, most admitted the task ahead of them would not be easy. Often they left behind young families to undertake a five-month voluntary placement delivering the polio vaccine. Moreover, they knew that they were putting their lives in jeopardy.

Selamawit Satato, a quietly-spoken 34-year-old nurse from the town of Soddo in Ethiopia, volunteered to spend five months working as a polio volunteer in northern Nigeria, where polio workers were targeted and killed in 2012 and where attacks by rebel groups like Boko Haram remain a serious threat. 'I feel I am in danger,' she admitted, 'but I want to contribute to eradicating polio . . . and even if there is danger I have decided to go to Nigeria in order to put all my efforts into eradicating this disease.'[5] Leaving behind her six-year-old son and two-year-old twins in the care of her mother, Selamawit said that she is sure that if she follows security instructions she will be safe but there are no guarantees. She knew that her son, in particular, was scared and unhappy that she was leaving: 'When I talked to him he was disappointed but I promised him that I will come back.'

Her colleague, a doctor from the Democratic Republic of Congo, who was about to embark on his second polio placement, explained the urgency and the pull, of being part of the final stages of the eradication campaign:

> We did not finish . . . we did not end this disease. I stopped to do other things but then I thought – no. I have to go on and I asked to return again so that we can work together until we end polio. The main thing is to be committed to something.[6]

Impossible idealism, intense commitment, immense difficulty, interminable political negotiations and sometimes bitter internecine rivalry have been the hallmarks of polio eradication from the very beginning.

A short and simple virus

Poliomyelitis, like smallpox, is caused by a virus (specifically, an RNA virus like influenza and hepatitis A). Unlike the smallpox variola virus, with its 200,000 nucleotides and 200 genes, the polio virus is short and simple, with a diameter of only three-millionths of a centimetre, 7,440 nucleotides and 11 genes.

The two are unalike in another crucial respect; while smallpox ravaged and killed millions, polio appears to have existed for centuries without causing great epidemics, surviving in a form that caused a mild infection in most people and paralysis in a few.

Records show polio cases in Ancient Greece and Rome and an Egyptian upright stone tablet from 1500 BCE clearly shows a young man with a withered leg. By the eighteenth century more frequent references to polio began to emerge. Sir Walter Scott wrote that when he was eighteen months old:

> One night . . . I showed great reluctance to be
> caught and put to bed . . . It was the last time
> I was to show much personal agility. In the
> morning I was discovered to be affected with
> fever. It held me three days. On the fourth . . .
> I had the lost the power of my right leg.[7]

A London doctor, Michael Underwood, wrote the first description of infantile paralysis following fever in 1789. By the 1800s, doctors were beginning to notice clusters of paralysed children in Europe and the USA and an Italian doctor, Giovanni Monteggia, recorded a similar illness in nursing babies. By 1840, a German doctor, Jakob von Heine, reported that he believed the disease was one that affected the spinal cord; this was borne out when the French neurologist, Jean-Martin Charcot, examined post-mortem material and located the disease to the grey central marrow of the spinal cord where motor function is located. Based on these findings, a German doctor, Adolph Kussmaul, coined the term 'poliomyelitis anterior cuta' from the Greek words *polios* (grey) and *myelos* (marrow).

Polio never carried a scourge of death, as did smallpox, cholera or measles or even influenza. Instead, its creeping mystery instilled a sense of terror in the western world, particularly in America. The first recorded polio epidemic in the USA was in 1894, when Charles Caverly, a young country doctor in Vermont, chased down and recorded 123 cases of an illness that had afflicted the children of Otter Valley near Rutland. Charles Caverly had a strong interest in public health and, although he had no idea what the illness was or how it spread, he understood that the disease had struck in summer, most of the victims were children and there was clearly a milder form of the disease that many more people contracted and recovered from without suffering paralysis. Of those 123 serious cases, 50 children were permanently paralysed and 18 died.

In 1905 a polio outbreak swept through Sweden, with 1,200 reported cases. A Stockholm paediatrician, Ivar Wickman, tracked the outbreak along railway lines and country roads, through rural towns and from school to school, before concluding that polio was a contagious illness, was capable of causing epidemics, and affected many people who did not realize that they were ill.

Three years later, in 1908, an Austrian doctor, Karl Landsteiner, made the first breakthrough in understanding the disease. Landsteiner had done pioneering work identifying different blood groups. Now, working with Erwin Popper, he transferred an emulsion of nerve tissue from a polio victim to a monkey and discovered that the disease could pass through a porcelain filter and therefore had to be a virus. Porcelain filters were a new and essential tool in microbiology, developed by Charles Chamberland, who worked with Louis Pasteur in Paris. These unglazed porcelain 'candles', with pore sizes of a ten-millionth of a millimetre, removed all bacteria from a liquid suspension but allowed minute viruses to pass through. Porcelain filters were instrumental in the birth of virology.

Karl Landsteiner's discovery that polio was a virus was a momentous leap forward in medical research but did nothing to explain where polio came from, how it spread or to slow the polio epidemics that were now breaking out every summer.

> In the summer of 1916, New York's playgrounds stood empty. No children splashed in public swimming pools; no one sold lemonade on the sidewalks. No cats roamed the alleys, peering into garbage cans. Troops of sanitary workers in white uniforms hosed down the city streets.

The city was in the grip of a major outbreak that would kill six thousand and paralyse 27,000 more:

> Fathers hurried home from work, fear imprinted on their faces, averting their glances from the tiny wooden caskets lined up outside the tenements. Policemen patrolled the streets. New York was a city under siege.[8]

Summer after summer the numbers of cases rose. Only six years later, all children leaving a once more polio-stricken New York had to be certified as free of the disease, while police in surrounding towns were instructed to turn back city-dwellers fleeing from the illness.

Polio outbreaks engendered widespread panic. Their tendency to cut across social and economic boundaries and pick off the children of wealthy families as easily as they preyed on the poor led to a growing urge to find out more and curb the disease. Building on Karl Landsteiner's work, the influential director of the Rockefeller Institute for Medical Research, Simon Flexner, replicated Landsteiner's observations and took them a step further, transferring the disease between monkeys and establishing an animal model for the disease. Cementing a thinking that dominated polio research for two decades, Flexner incorrectly concluded that polio was a disease of the nervous system that passed through the nasal nerves to the brain and spinal cord. Convinced that a vaccine would be ineffective, Flexner devoted his attention to preventing infection by means of a nasal blockade and attempting to stop the spread of the disease with chemical nasal sprays. From 1936 he co-opted Peter Olitsky to work on the disease, together with his young associate, Albert Sabin.

Although polio notably struck at the young it was by no means a disease that only affected children. The response of one adult victim who contracted the disease during the outbreaks of this era shaped the course of polio research for decades.

The March of Dimes

In the summer of 1921 Franklin Delano Roosevelt was taking a holiday from his demanding job in Washington DC. Exhausted by a gruelling and unsuccessful campaign for the vice-presidency the previous year, Roosevelt was further beset by a scandal that had engulfed him in his role as assistant secretary to the Navy; it was claimed that he had endorsed a secret plan to uncover evidence of homosexuality by using undercover agents to entrap sailors at a naval training centre.

When, finally, Roosevelt felt able to leave Washington DC, he stopped off briefly at a Boy Scout jamboree at his home in Hyde

Park, before joining his family on Campobello Island, off the coast of New Brunswick and Maine. The day after his arrival, on 8 August, he accepted a challenge from his children and went sailing, swimming and running, before spending the afternoon answering letters, still wearing his wet bathing suit. As the afternoon wore on Roosevelt noted that he'd 'never felt quite that way before', complaining of flu-like symptoms, muscle aches and a fever before retreating to bed for the evening. When he woke the next morning and pulled himself out of bed for a shave, he noticed that his left leg was dragging. A doctor was summoned, who reassured him that he was probably suffering from nothing worse than a 'bug'. However, his condition worsened, until he had lost the use of both legs and his skin became so sensitive he could hardly bear the feeling of his pyjamas. Lying in bed, muttering over and over, 'I don't know what's wrong with me,' Roosevelt had, at the age of thirty-nine, become the most prominent American to be felled by poliomyelitis.

Although photos of Roosevelt in his wheelchair rarely appeared during his lifetime (after his election as president the Secret Service ripped the film out of the cameras of anyone they saw attempting such a shot) the American public was well aware that their leader suffered from a crippling disease. News of his illness was first reported on the front page of the *New York Times* and a 1932 *New York Times Magazine* profile when he was governor of New York described how he 'wheels around in his chair'. Another article that year in *Time* magazine said that swimming and exercise 'have made it possible for the Governor to walk 100 feet or so with braces and canes. When standing at crowded public functions, he still clings precariously to a friend's arm.'[9]

It was not that voters did not know that Roosevelt was disabled, but that they believed he had suffered from, and triumphed over, polio. Roosevelt began to extend his efforts to fellow polio sufferers in the mid-1920s, realizing that others, like him, could benefit from the recuperation and hydrotherapy on offer at the ramshackle Meriwether Inn and cabins in Warm Springs, Georgia. Although his wife Eleanor disliked the racism and rural life of the area, describing it as 'hard and poor and ugly', F.D.R. believed his stay there was key to his recovery. When other polio sufferers were shunned by guests who worried that they too would catch the disease, Roosevelt purchased the resort. After his election to the presidency in 1932, he left the resort in the hands of his young law partner, Basil O'Connor.

O'Connor was a purposeful, if rather charmless, self-made man who had risen from working-class Irish roots to earn a place at Harvard, where he studied so hard he went temporarily blind. In his new role at the Warm Springs Foundation he began an intense fundraising campaign to keep the charity afloat through the worst days of the Great Depression. He was not sentimental about his decision to step in, later saying that 'My decision had no more emotional significance than taking over several file folders of unfinished business for a colleague.' Under O'Connor's leadership the foundation became the biggest voluntary health organization in America, mobilizing support and harnessing the power of radio and film to reach out to every community in the country.

In the immediate aftermath of the Wall Street Crash in 1929 the Warm Springs Foundation almost collapsed, as donations from wealthy donors dried up overnight, but O'Connor called on a former insurance salesman, Keith Morgan, and a public relations guru, Carl Byoir, to 'sell' the concept of Warm Springs and change attitudes towards what was often thought of as a mysterious childhood disease. Byoir suggested a nationwide fundraising ball to celebrate Roosevelt's birthday and employed every technique he had learned in advertising and PR to make it happen. On the night of 29 January 1934 six thousand local fundraising balls took place across the nation, spearheaded by an evening at the Waldorf Astoria in New York where five thousand guests watched more than fifty debutantes mount a multi-tiered birthday cake twenty-eight feet in diameter. With the final contributions totalling more than one million dollars, Roosevelt announced that it was the 'happiest birthday I have ever known'.

The idea that the generosity of ordinary Americans, donating small sums, would support polio victims and fund a medical breakthrough was summed up by the singer Eddie Cantor, when he suggested to a meeting of Hollywood moguls that 'the March of Dimes' should be the slogan for a radio campaign in support of the official launch of the newly formed National Foundation for Infantile Paralysis (NFIP), which Roosevelt had announced in September 1937. The foundation was an attempt by Roosevelt to de-politicize the polio campaign amid accusations that it was a Democrat-backed movement. Again, it was Basil O'Connor whom he asked to take the helm.

Explaining the rationale behind the slogan, Eddie Cantor said:

> The March of Dimes will enable all persons, even the children, to show our President that they are with him in this battle against this disease. Nearly everyone can send in a dime, or several dimes. However, it takes only ten dimes to make a dollar and if a million people send only one dime, the total will be $100,000.[10]

The first two days of the campaign brought in a meagre $17.50 and Cantor's pitch seemed to have fallen flat. But soon the money and letters of support began to roll in, until by the end of the first phase of the campaign Roosevelt had received eighty thousand pieces of correspondence and $268,000.

On the eve of his birthday ball, President Roosevelt responded:

> During the past few days bags of mail have been coming, literally by the truck load, to the White House. Yesterday between forty and fifty thousand letters came to the mail room of the White House. Today an even greater number – how many I cannot tell you, for we can only estimate the actual count by counting the mail bags. In all the envelopes are dimes and quarters and even dollar bills – gifts from grownups and children – mostly from children who want to help other children to get well.[11]

Entertainers including Ginger Rogers, Jean Harlow, Bing Crosby, the Lone Ranger, Mickey Rooney and Judy Garland all supported the March of Dimes campaign, encouraging people to play their small part in tackling a disease that was believed by some to pose as great a national threat as a world war. Indeed, at its height, the National Foundation for Infantile Paralysis operated like an army, sending ninety thousand year-round and two million seasonal volunteers into battle against a disease that – according to the NFIP's publicity film – crept across the sky like a dark cloud, intoning 'My name is Virus Poliomyelitis . . . I specialize in grotesques, twisting and deforming human bodies, that's why I'm called "The Crippler".'[12] Some did not join the march. Objections focused on the immense fundraising efforts directed *exclusively* to a disease that claimed far fewer victims than many others. Ultimately, however, such voices were drowned out by the sinister whisper that continued: 'I am very fond of children, especially *little* children.'

While Americans feared polio second only to a nuclear attack, Britain, by contrast, responded to outbreaks with calls to keep calm and remain optimistic. A UK Ministry of Health film, released in response to a 1947 epidemic that paralysed 7,800 and killed 700, featured the two-year-old Johnny Green, who stared benignly into the camera, unable to raise his hands or head off the mattress. 'Go ahead Johnny and win through,' urged a stirring voice-over from Eleanor Roosevelt.

Crucially, however, the March of Dimes not only mobilized a response to a terrifying threat but generated funding to support research: 'Like the sudden appearance of a fairy godmother of quite mammoth proportions,' as the polio researcher John Paul described it. But in this first era of vaccine research the results were haphazard, inconclusive and sometimes devastating. One terrible year, 1935, became the year of the failed vaccines.

The most notable failure was the Brodie-Park vaccine, developed by Maurice Brodie, a young Canadian researcher at the New York University School of Medicine, who worked with a rival of Simon Flexner, William H. Park. Together Park and Brodie developed a vaccine that, using monkeys, they proved produced antibodies.

The two men then injected themselves, with only minor discomfort. After vaccinating a dozen children, Brodie and Park published their study in the *Journal of American Medicine*. On this rather slim basis some states went ahead and permitted trials but when larger numbers of children were inoculated the vaccine did not produce a sufficient number of antibodies and sometimes provoked severe allergic reactions. Brodie's career was blighted; he died at thirty-six, probably by suicide. Another researcher, John Kolmer, who

> Vaccines work by imitating an infection, stimulating the immune system into producing T-lymphocyte cells and antibodies. Once the imitation infection caused by the vaccine (which does not itself provoke serious illness) goes away the body is left with a 'memory' of T-lymphocytes and B-lymphocytes – T-cells and B-cells – that 'remember' how to fight the real disease in the future.

was based in Philadelphia, produced a rival vaccine that he tested on himself, his two sons and twenty-three other children. He soon broadened his trial to ten thousand children but with devastating results: nine children died of polio attributable to the vaccine.

These vaccines were real and serious failures for polio research and it would be another twenty years before the race to produce a viable vaccine started again. However, one interesting result emerged from the research of 1935: proof that Simon Flexner's theory on how polio entered the body was wrong. Flexner had exclusively relied on the monkey model of polio but the species he used, the rhesus monkey, was not susceptible to polio by the natural route of infection. Rhesus monkeys can only catch polio either through the nasal nerves or injection into the brain but in humans, polio is a gastrointestinal virus that replicates in the cell lining of the gut and is passed from person to person through faeces and contaminated food and water. Flexner's results were affected by his creation of a strain of the virus that was restricted to nerve tissue, and that strain was sent to many research laboratories throughout America, misleading many researchers and setting back progress towards finding a vaccine.

In the face of these setbacks, the NFIP deemed that a period of meticulous research and fact-finding was needed. Basil O'Connor appointed Thomas Rivers, director of the Rockefeller Institute Hospital, to lead a committee on scientific research. Rivers drew up a list of topics he believed needed to be addressed – including how the virus gets into the body and how it is transmitted – and selected scientists from Yale, the University of Michigan and Johns Hopkins University to work on this research.

By the end of the 1930s this new research was under way, while the fundraising efforts of the March of Dimes continued to raise unprecedented amounts of money, unstaunched even by the Japanese attack on Pearl Harbor and the entry of the USA into World War Two. In 1938 the March of Dimes collected $1.8 million, rising to an astonishing $19 million in 1945. The march of polio, however, remained relentlessly grim, with 9,000 cases in 1941, soaring to 20,000 by the end of 1945 and more than 25,000 in 1946. Furthermore, the NFIP lost its figurehead and inspiration in April 1945, when Franklin Delano Roosevelt died of a cerebral haemorrhage at his 'Little White House' in Warm Springs, Georgia. The cloud of 'The Crippler' hung ever more ominously over every home.

With its new emphasis on research, the NFIP appointed Harry Weaver as research director in 1946. Weaver framed an additional set of primary questions that needed to be answered, including determining how many strains of polio existed. Two were known, but to develop an effective vaccine researchers needed to be sure they knew all the types. In 1949, David Bodian and his team at Johns Hopkins defined three types of polio virus and Weaver began the funding of what would become a five-year, million-dollar programme to ascertain whether these three serotypes (variants of the virus) represented the full extent of the virus. One of the researchers who worked on this long project was Jonas Salk, at the Pittsburgh University School of Medicine.

In addition to the serotype project, the possibility of producing a vaccine was becoming more feasible. Frederick Robbins, John Enders and Thomas Weller, from the Boston Children's Hospital, made a crucial breakthrough (which would eventually win them the Nobel Prize in Physiology and Medicine), successfully cultivating the polio virus in non-nervous tissue culture, including skin and muscle cells. This made it possible to grow large quantities of tissue cultures to produce a vaccine.

Albert Sabin and Peter Olitsky had demonstrated that poliovirus could grow in human embryonic brain tissue in 1936. The advantage of embryonic brain tissue was that it grew quickly but using that tissue could provoke central nervous system damage if the cultivated virus was then used in humans. Enders, Weller and Robbins's discovery used human embryonic skin and muscle tissue mixed with antibiotics, which grew quickly but was safe to develop as a human vaccine. A second major discovery came as the result of research first by Albert Sabin in 1941 and then by David Bodian and Howard Howe at Johns Hopkins, that used experiments on chimpanzees to confirm that polio was a disease of the gastrointestinal tract.

It seemed that polio research was developing along a well-planned path but, much to the astonishment of researchers, the development of a polio vaccine was about to lurch in a totally unexpected direction.

The close-knit polio research community was rocked when, in March 1951, a Polish refugee, Hilary Koprowski, announced that he had secretly developed 'The first successful trial of immunization of man against poliomyelitis.'[13] As far as other researchers were concerned, it was news of a vaccine from a source they had never imagined. Koprowski graduated from Warsaw University Medical

School before fleeing Nazi persecution. As well as his medical career, he was an accomplished musician and composer. He had begun work with the pharmaceutical company Lederle in New York State in 1945; within two years he was injecting type 2 polio virus into the brains of mice and then passing the virus through several groups of animals until he had produced a weakened strain, which he finally fed to nine chimpanzees, all of which were protected from polio. Koprowski moved on to testing in humans by drinking the vaccine himself, straight from his kitchen blender. The *New York Times* wrote:

> The main ingredients were rat brain and a fearsome, carefully cultivated virus. In his laboratory in Pearl River, NY, twenty miles north of Manhattan, Dr Hilary Koprowski macerated the ingredients in an ordinary kitchen blender one January day in 1948. He poured the result – thick, cold, grey and greasy – into a beaker, lifted it to his lips and drank. It tasted, he later said, like cod liver oil.[14]

Koprowski began testing his vaccine on children, working with his friend George Jervis, who was laboratory director at the Letchworth Village, a state institution for 'feeble-minded children' in the Hudson Valley. They tested twenty Letchworth children with no ill effects. Koprowski prepared to astound the polio community by announcing his results at a NFIP round table meeting in Hershey, Pennsylvania, attended by Albert Sabin, Jonas Salk, David Bodian and Thomas Rivers. 'The data I want to acquaint you with,' Koprowski began, 'represent a summary of clinical trials based on oral feeding of the TN [Thomas Norton] strain of polio [living virus].'[15]

After the failed vaccines and deaths of the 1930s, Albert Sabin, in particular, was outraged at the idea of testing a live attenuated (weakened) virus on a human being, repeating 'Why did you do it? Why?' Koprowski insisted that 'somebody had to take the next step' and that his work was in line with the accepted practice of the time. He went on to produce an attenuated strain 1 polio vaccine by the same route, this time using the breakthrough in tissue culture to produce his material. Despite losing his job at Lederle in 1954, Koprowski continued to test his vaccine throughout the 1950s and inoculated more than a quarter of a million people in what was then the Belgian Congo. A 1958 trial in conjunction with Queens University, Belfast was halted, however, when stool samples from vaccinated children revealed the virus had reverted to virulence. It seemed that Koprowski's vaccine had not remained sufficiently weakened.

'What you have to understand about Koprowski was that he was *the* leading polio researcher from private industry', wrote Professor David M. Oshinsky. 'Had he come out of the university and had he been funded by the March of Dimes, I think he would have been a favourite of the virology community.'[16]

Koprowski went on to head the Wistar Institute in Philadelphia, where he developed a new rabies vaccine and oversaw the development of the rubella vaccine in the 1960s. Yet, according to Oshinksy, 'There was a great deal of hurt and jealousy on his part that he never got credit for what he deserved and for the path that he blazed for others to follow.'[17] Jonas Salk and Albert Sabin were about to slug it out in the battle to vaccinate the world against polio; Koprowski languished, a forgotten man.

Salk vs Sabin

There were others in the running but Salk and Sabin became the titans of polio research, locked in a battle between their two personalities and two different concepts for creating a polio vaccine. Based on his research on a flu vaccine, Salk was convinced that a *killed* polio vaccine would be effective. Sabin was equally adamant that only a *live* vaccine was the solution. And their battle was personal; Sabin descended to a level of spite that would forever colour the reaction to his own significant achievements.

Jonas Salk was a relative newcomer to the polio research community. Born in New York in 1914, he was the son of Jewish immigrants from Russia who worked in the garment industry. Salk was the oldest son of the family and grew up in the shadow of his domineering mother. He was, by all accounts, a shy, serious boy. According to his son, Peter: 'He was basically born an adult . . . he didn't have the freedom to be a child in his family.' This trait, combined with an agreeable manner that often masked deep inner turmoil, shaped his life.

Salk studied hard and was the first in his family to go to university. He attended Townsend Harris, a school for gifted children, before entering City College, initially as a law student. Although he worked hard, Salk's grades were unremarkable and his mother persuaded him to switch to pre-medicine. Salk admitted that if he could not win an argument against his mother he was unlikely to become a successful attorney. After changing to pre-medicine he improved his grades just enough to get into New York University College of Medicine, where he excelled, and was soon asked to become involved in his first research projects. Crucially, at NYU Salk began studying under the man who became his mentor, Thomas Francis, a professor in the bacteriology department who was working on an influenza vaccine.

In a trait that would become apparent throughout his scientific career, Salk marked himself out as a pleasant and congenial outsider, but one whose personality initially made little impact. One boy who sat next to him in class for a year barely remembered him, while at Townsend Harris he was overlooked for election to an élite student society many believed he should have been a natural contender for. 'Jonas was not very verbose,' Walter Kees told his biographer Charlotte DeCroes Jacobs; he was 'not a social man'.[18]

Salk graduated in 1939 and two days later married Donna Lindsay, a cultured, educated and glamorous girl far above him on the social scale, whom he had met during a summer job at the marine life laboratories in Woods Hole, Massachusetts. His fiancée's family was less than impressed with the match, which they considered a step back into the immigrant world from which they had only recently emerged. Donna's father insisted on two conditions before agreeing to the union. One was that Salk should qualify as a doctor before they married; the second was that he should add Edward as a middle name, to sound more impressive. (This obviously rankled; when the couple divorced one of Salk's first acts was to revert to plain Jonas Salk.) In hindsight, Woods Hole was significant in Salk's life for another reason; there, he met his great rival and nemesis, Albert Sabin, for the first time.

On leaving NYU, Salk took up a place at Mount Sinai Hospital in New York and, in addition to his work, became involved in left-wing politics, campaigning against fascism in Europe and inequality in New York. This was an interest he initially shared with his wife Donna but while she remained committed to such causes, Salk became increasingly apolitical, as his career consumed his attention. What remained, and guided him throughout his life, was a

sense of idealism and a desire to do good works, in the Jewish tradition of *tikkun olam* ('world repair'). Salk himself said: 'I became
aware of a desire to do something in life that would relieve some of
the suffering'. Showing less reverence, his younger brothers just
called him 'little Jesus'.[19]

After completing his training, Salk's preference was to stay at
Mount Sinai or join the Rockefeller Institute but those jobs were
not offered him. He wrote to his mentor from NYU, Thomas
Francis, who was by then chairman of epidemiology at the
University of Michigan School of Public Health at Ann Arbor. Two
hundred years earlier, Edward Jenner's breakthrough with the
smallpox vaccine had established that live, attenuated, vaccines
were the most effective. Now, established figures in the polio
community, including John Enders and Albert Sabin, believed that
a live polio vaccine was the *only* solution. In contrast, Thomas
Francis, the first scientist to isolate a human influenza virus, who
went on to identify Type A and Type B flu, was an expert on inactivated vaccines and on the verge of developing a killed flu vaccine
for the US Army.

Francis, a 'short, pudgy, somewhat proper man',[20] could be
'fussy' and 'pitilessly critical' of his students, although he often
formed lifelong bonds with them. In responding to Salk's requests
for work, he did little to encourage him, replying to his letters with
calls of restraint, asking him to 'cool off' before making any drastic
decisions. Salk eventually arrived in Ann Arbor on 5 March 1942,
for an unsalaried research appointment in the Department of
Epidemiology. He did, however, have his first NFIP grant for
immunology research.

Over the course of six years Salk learned much from Francis,
taking the lead in the country's largest trial of a flu vaccine (something of an urgent priority after the terrible pandemic of 1918–19,

which killed more people than World War One) and proving that a killed vaccine could be effective. Although in those years Salk developed many of the skills and techniques that would serve him for the rest of his career, it was clear that he liked to proceed in leaps rather than the incremental, meticulously researched steps advocated by Francis, who offered a useful restraint to some of his protégé's impulses.

When Salk formed relationships with the press and pharmaceutical companies, Francis was outraged. Furthermore, Salk tended to write long rambling research papers, completely at odds with the style of manuscripts published by medical journals. After reading one draft paper, Thomas Francis informed Salk that he did not have enough data to substantiate his conclusion. Salk said that the conclusions were warranted by 'reason' rather than 'hard data' and he intended to submit his paper for publication. Francis replied that if he did submit, he should leave with it. Salk did not leave immediately but fault lines were emerging in his relationship with Francis. Thomas Francis was a valuable ally and mentor at a crucial period in Salk's career but he had discovered a weakness that Salk's future enemies would exploit. Salk's sloppiness in presenting his work was a trait that ensured he remained an outcast from the scientific community. As he admitted, ever since he had chafed against his overbearing mother, he had wanted to do things his own way.

By 1947 Salk was ready to leave Ann Arbor and Thomas Francis, surprising his colleagues by taking a position at the Medical School in Pittsburgh, not an institution renowned for its research. Salk began his tenure in a run-down, disused basement that had previously housed the morgue. But while it was far from a glamorous position, it provided Salk with the opportunity to be his own man, build a research team and conduct the work he felt compelled to do. Salk believed the main thrust of his efforts would be spent on developing an influenza vaccine, but in 1948 he received the first funding from Harry Weaver, the NFIP's director of research, to work on the different strains of polio. Although Salk had not previously worked on polio, Weaver had noticed his impressive work on influenza. Of equal importance was the fact that while many scientists did not want the NFIP to guide their research, Salk was delighted to take the foundation's funding, believing that it would amply support his influenza project. Salk eventually had to choose between influenza and polio; as the years passed, Salk's funding for polio serotyping increased, until his laboratory accounted for the vast majority of the external funding at Pittsburgh Medical School and his department occupied two full floors of the building.

The success of the typing project meant that by its conclusion in 1951 Salk was a middle-class professor, running a successful laboratory, with three sons and a house in the suburbs. It was not a project that had brought him great esteem, however. In the eyes of others within the polio community, Salk was a lowly researcher, carrying out the most mundane kind of work. It was an opinion that would stick when Salk progressed to a far loftier ambition; the development of an effective polio vaccine.

1952, the year that Dwight Eisenhower won the presidency, was a year of interesting medical advances, including the first successful surgical separation of conjoined (Siamese) twins and in Denmark, the world's first sex change surgery. But it was also a terrible year for polio in America, with nearly 58,000 reported cases, 21,269 people left paralysed and 3,145 dead. Jonas Salk's first human trials of a polio vaccine could not have been more timely.

Salk had tested both a live attenuated and a killed vaccine two years earlier in monkeys. Using a formaldehyde inactivation developed for the flu vaccine to kill the toxins, Salk found that the killed polio vaccine was effective and stimulated a good level of antibodies. When the time came to approach the NFIP, Salk found Harry Weaver cautious but supportive. He had informed Weaver as early as 1948 that he planned to have a vaccine in five years. 'There was no one like him in those days,' Weaver said. 'He thought big . . . he was out of phase with the tradition of narrowing research down to one or two details and making progress inch by inch. He wanted to leap, not crawl.'

Salk won over the NFIP chief Basil O'Connor as an important ally, forming a good relationship with him when they travelled back from Copenhagen on the *Queen Mary* in September 1951, after Salk had successfully presented his results from the typing project to the Second International Poliomyelitis Conference. Salk said he had started talking to O'Connor as they waded waist-deep in the ship's swimming pool: 'He made me feel as if I could see more broadly, more clearly, more deeply then I could when alone or with others . . . I felt I had found a kindred soul.'[21] Three months later, in December 1951, Salk's proposal to hold a human trial went before an expert committee comprising Weaver, Albert Sabin, David Bodian, John Enders, Thomas Francis, John Paul, Tom Rivers and Salk himself.

Salk believed he had the tools to produce a vaccine. Two major advances had been made since the failed vaccines of the past and the Koprowski vaccine: Salk had identified all three strains of polio and had mastered viral growth in tissue culture, instead of using rat brains or nervous tissue. Despite this, Enders and Sabin were implacably opposed to Salk's proposal, believing that only a live attenuated virus would be successful. The immense success of the March of Dimes also brought immense pressure; the NFIP had been promising a breakthrough polio vaccine since 1949 but there was still nothing to show. When the committee rejected Salk's proposal Harry Weaver felt he had only one course of action: ignore their decision. Weaver consented to move forward, in secret, with Salk's trial. Testing began in June 1952 at the D. T. Watson Home for Crippled Children and the Polk School for the Retarded and Feeble-Minded.

'This is it!' Salk proclaimed to patients and staff at the Watson Home on 2 July 1952, holding up a small vial of his vaccine for all to see. The first volunteer was sixteen-year-old Bill Kirkpatrick, who had been preparing to play high school football when struck down with polio. Kirkpatrick said the disease was like 'someone getting a sledgehammer and beating against your spine. I felt my legs get soft like jelly.'[22] Salk chose someone who had already caught polio because the trial was to prove the safety of the vaccine and its ability to stimulate the body into producing antibodies, not to prevent the disease.

Salk, who had a kind and attentive bedside manner, inoculated every child himself and waited anxiously for the results, later telling a reporter, 'when you inoculate children with a polio vaccine for the first time you don't sleep well for two or three weeks'. Much to Salk's delight the vaccine was safe and the children tested produced good levels of antibodies.

However, the announcement of the trial results at a meeting of the NFIP Immunization Committee in Hershey, Pennsylvania on 23 January 1953 met a stormy response. Albert Sabin was astonished that Salk had welcomed him for dinner in his house the previous evening and travelled with him on the train to Hershey without once mentioning the trial. Although the results were momentous, they were rather disingenuously listed as item number two on the agenda: 'A presentation by Jonas Salk of some recent work'. Salk first acknowledged the work of Isabel Morgan (a colleague of Bodian and Howe) who had vaccinated monkeys with inactivated polio virus, then took everyone by surprise by announcing that he had tested his own vaccine on fifty-two children at the Watson Home.

Sabin, Enders and Howe remained opposed to a further large-scale trial, claiming the use of the virulent Mahoney strain of the virus, as well as issues about a mineral oil adjuvant (a substance added to vaccines to increase the body's immune response), raised safety concerns. In a manner that Salk was now painfully used to, Sabin set about dissecting the research with an imperious sense of dismissal: 'We hadn't done this or we hadn't done that,' Salk told Richard Carter. 'His interpretations made my work seem incredible, of no meaning or significance.'[23]

Salk was unsure about proceeding with a large-scale trial, believing it to be premature, but Harry Weaver and the NFIP were impatient to move forwards. When an NFIP source leaked the news, the *Pittsburgh Press* reported that Salk and his colleagues had been working on the vaccine that could 'spell final doom for polio'. Placed in the media spotlight, Salk appeared on a CBS television special to explain that although 'certain things could not be hastened', there was 'justification for optimism', and he promised to publish an article on his work in the *Journal of the American Medical Association* on 28 March. When the article was published, journalists took it as the proof they were looking for, ignoring the researchers Salk listed as his colleagues and turning his tentative conclusions into definite statements. Salk's peers were less impressed by the somewhat rambling paper, which had undergone minimal editing, skipping back and forth between human and animal testing and mixed methods, data and conclusions. It was the epitome of the sloppy work Thomas Francis had warned Salk about years earlier.

The NFIP decided to march ahead regardless, pushing aside Salk's worries and the opposition of its own immunization experts. Basil O'Connor convened a new committee to design a large-scale trial, which included no members of the existing immunization committee. The design of the trial would once more set the NFIP against the scientific community. Basil O'Connor and Salk argued for moving speedily forward with an uncomplicated trial but Joseph Bell, who was appointed to design the trial, vehemently disagreed. Bell argued for the far more rigorous approach of a 'double blind' trial and control groups, the standard for all drug trials today.

A stand-off ensued, with heated and emotional negotiations that pitted O'Connor against Harry Weaver and ended with Weaver's resignation. Joseph Bell's proposal for a double blind trial was rejected by the committee, and Bell resigned on 1 October. Salk's mentor, Thomas Francis, agreed to take over the trial, on the condition that he could operate with complete autonomy. That demand granted, Francis reinstated the double blind element of the trial, insisting that only the most rigorous approach would suffice.

The scale of the trial was immense: tens of thousands of professionals and volunteers across America immunized 600,000 children between the ages of seven and eight with three separate injections administered over a period of eight weeks. Amid a mood of national excitement, fears and worries emerged. Would the pharmaceutical companies Eli Lilly and Park Davis be able to produce enough vaccine? They could, but others, including Cutter Laboratories, were added as backups. Was including the Mahoney strain of the virus safe? Would inactivation work? On the eve of the trial the press ran a damaging story that live polio virus had got through the process and caused the disease in monkeys. David Bodian was called in to investigate and triple safety checks were added to make sure that no defective vaccine got through.

Finally, on 26 April 1954, Randy Kerr, a six-year-old student at Franklin Sherman Elementary School in McLean, Virginia, stepped forward to receive the first shot, administered by a local doctor, Richard Mulvaney. The school in McLean secured its place in history by accident. The trial was supposed to be held in Washington DC, but the city pulled out after the broadcaster Walter Winchell made a scaremongering radio address about the vaccine's safety. Mulvaney remembered:

> When I walked into where the vaccines were to be given, there were quite a lot of people. There were a lot of reporters, TV cameras and kids out in the hall screaming. They were screaming because they were about to get a shot. I wasn't prepared for the reaction.[24]

Gail Adams Batt received the vaccine at Franklin Sherman Elementary School, even though she attended another local school: 'It seemed like a fun adventure. There were three of us in our little group – I was the youngest and was proud to be included. It did seem like we were doing something special at the time.' Batt always wondered if she had been chosen to make the trip over to Franklin Sherman Elementary because she sat next to a boy in her normal class who had recently come down with polio. 'The day was a success in my mind. I did not cry, I did not faint . . . and I got to leave school and get an ice cream,' Batt said.[25]

By the end of June all the vaccinations were completed. Over the summer, America waited to see if the vaccine would work. And wait it did, for more than nine long months, until 12 April 1955 when Thomas Francis presented his 563-page report on the trial at the Horace H. Rackham Auditorium in Ann Arbor, Michigan.

The day dawned bleak and cloudy. Salk, who was staying at a university residence with his wife and children, ate a nervous breakfast with Basil O'Connor and others involved in the trial. Although he was confident about his work Salk remained anxious about various aspects of the trial that had been taken out of his control: the introduction of a chemical called merthiolate and the fact that the third booster shot had been given to children weeks earlier than planned, to coincide with the end of the polio season. None of the men present knew the result until Thomas Francis arrived and told them the verdict: 'The vaccine works. It is safe, effective and potent.'

The atmosphere inside Rackham Hall was feverish. One hundred and fifty members of the press had assembled from around the world, served by forty-one newly installed telephone lines and six teletype machines. Pacing like 'hungry dogs', the reporters restlessly circled the library tables, waiting for the University of Michigan press office to hand over copies of the report, promised for precisely 9:10a.m. The scene descended into an unruly mob, with journalists clamouring for the hastily tossed-out packages, which arrived late, at 9:17a.m. Down below, in front of twelve hundred guests, a soberly dressed Thomas Francis began an hour-long, scientifically accurate, explanation of his findings. Upstairs the press room exploded: 'It works!' shouted one reporter after the barest glimpse, immediately breaking the gentlemen's agreement not to publish the conclusions until Francis had finished speaking. NBC's *Today* broke the news to the American people first, admitting it was just too good to wait.

Overall, the vaccine was eighty to ninety per cent effective against poliomyelitis, Thomas Francis concluded. It had proved to be ninety per cent effective against Types 2 and 3 and sixty to seventy per cent effective against Type 1. 'POLIO ROUTED' exclaimed the *New York Post*. Church bells rang, car horns blared and shop owners wrote 'Thank you Dr Salk!' across their plate glass windows. The nation burst out in a mass expression of jubilation not seen since VJ Day. After waiting so long, the clamour for the vaccine was immediate; by mid-April six commercial manufacturers had enough supply to begin distribution.

If the first leaked news of Salk's trial subjected him to intense media scrutiny, the success of the vaccine raised him to the status of American icon. Salk's wife Donna reflected ruefully on the fact that they left their home in Pittsburgh as a relatively anonymous American family and returned in a motorcade with a police escort: 'The world had changed and I must say from our point of view not for the better.' The Salks and their sons were swamped by a deluge of requests for interviews and photo shoots and buried under bulging mailbags of letters from grateful parents, polio victims and ordinary citizens who appreciated the calm humility of the hero as presented to them by the media. Salk's five-year-old son Jonathan summarized their new situation when he told his friend: 'I'm back from my vacation and I'm famous and so is my Dad.'[26]

Salk later faced criticism from his research team at the University of Pittsburgh that he had not given them full credit for their contribution, but in the eyes of the American people Salk could do no wrong. Admiration for Salk reached new heights when, appearing on *See It Now* with Ed Murrow, Salk announced that the 'people' owned the patent, adding: 'The vaccine has no patent. Could you patent the sun?' This statement was in keeping with Salk's much-stated desire to do something good for humankind. Salk resolutely refused to profit financially from his breakthrough, turning down all offers of financial reimbursement and continuing to live on his university salary. Adored as he was, however, Salk had some very dark days ahead.

On 24 April, a little girl in Idaho, who had received Salk's vaccine six days earlier, came down with the symptoms of polio. Three days later she died. In that short space of time four more polio cases were reported in that state. Soon the number of polio cases rose to a dozen, spread from Chicago to San Diego. The common factor was that each had received a vaccine made by Cutter Laboratories in Berkeley, California. On 27 April the US surgeon general halted use of the Cutter vaccine with immediate effect and on 8 May, ordered a stop of all use of the vaccine until it could be proven safe. By the end of the summer of 1955 two hundred polio cases, with eleven deaths, had been attributed to defective vaccines from Cutter laboratories. Later studies demonstrated that Salk had underestimated the complexity of the formaldehyde inactivation process and, at the Cutter laboratories, live virus had withstood the process designed to kill it. Cutter's use of glass filters (rather than the asbestos filters used by Salk) meant that the virus survived by clinging to bits of cellular material, and making into the final vaccine.

Salk was suicidal, besieged by criticism and distraught by the shilly-shallying of the US surgeon general, who seemed unsure about the correct course of action to remedy the crisis. Once the situation was resolved, Salk's vaccine proved safe and polio cases fell from 35,000 in 1953 to 161 by 1961. It was an incredible success, but it came too late for Salk. By then, his vaccine had been replaced in the USA by one made by his rival and harshest opponent, Albert Sabin. Sabin had fought Salk every step of the way and eventually, he scented victory.

Although he wore a tweed coat and affected an imperious elder-statesman manner, Sabin was only eight years older than Salk. Born Abram Sapersztejn in what was then the Russian city of Bialystok, his Jewish family escaped the Russian pogroms and emigrated to the United States in 1920, settling in New Jersey.

Recognizing his intelligence and abilities, Sapersztejn's uncle offered to fund his studies at dental school but his nephew soon tired of dental work and began a career in medicine and research instead, changing his name to Albert Sabin when he applied for US citizenship.

Unlike Salk, Sabin had an uneasy bedside manner and preferred working in the laboratory to treating patients. Inspired by Paul de Kruif's book *The Microbe Hunters*, the young Sabin successfully lobbied William H. Park, who was then dean of the Medical School at New York University, to take him on at NYU and help him find casual jobs to pay for his studies. Sabin proved an excellent researcher, publishing his first paper on the physical properties of the gels used to grow bacterial cultures in only his second year of medical school. In 1931 Sabin delayed a residency at Bellevue Hospital to begin work on a polio project. Within four years he began work at the Rockefeller Institute with Peter Olitsky, where he published forty papers in four years and disproved Simon Flexner's idea that polio was a disease inhaled through the nose; something that left Flexner 'not very happy', Sabin recalled.

In 1939 Sabin turned down an offer to work with the world-renowned virologist Sven Gard, at the Karolinska Institute in Stockholm, opting instead to become an associate professor of paediatrics in Cincinnati. 'An intriguing job title for a man who actively disliked children and whose main clinical experience (four thousand post mortems) had provided rather limited opportunities for honing his communication skills,' noted Gareth Williams in his book *Paralysed With Fear*.[27]

Sabin gained a stellar reputation in polio research and, after war service as a civilian adviser on viral infections, he first proposed creating an oral polio vaccine in a letter to the NFIP in 1946. Funding was not forthcoming until mid-1952, by which time

Sabin had a lot of catching up to do. His attack was two-pronged: 'One prong was an intensive laboratory programme to find and attenuate suitable strains of polio virus; the other was a vicious campaign to sabotage his competitors. His main target was Salk.'[28]

As a long-established scientist, Sabin looked down on Salk and spoke out against the clinical trials of Salk's vaccine at every turn. Taking advantage of the disaster at Cutter laboratories, Sabin stepped forward to state that now was the time to develop his live attenuated vaccine. Working from his laboratory at the Cincinnati Children's Hospital, Sabin successfully tested his triple-vaccine on thirty prisoners at the Chillicothe Federal Penitentiary in Ohio. The results were successful and because the attenuated virus grew naturally it stimulated a better response in the immune system than Salk's inactivated virus.

With Salk's vaccine already in place, however, the NFIP had little desire to test Sabin's vaccine. Sabin would find that his vaccine proved to be most successful in a vast immunization programme undertaken by the Soviet Union. Despite their interest in his vaccine, Jonas Salk had regretfully turned down an opportunity to visit the Soviet Union, not for ideological reasons but because his wife Donna insisted he spend time at home with his family. It was a costly decision for Salk, as Sabin stepped in, striking up a strong relationship with Mikhail Chumakov at the Moscow Polio Research Institute, who agreed to test the Sabin vaccine on ten million children in 1959. When the results were successful, the Soviet Union ordered a mass immunization campaign for seventy million children, verified by the WHO.

At the urging of the *Journal of the American Medical Association*, trials of the Sabin vaccine began in Ohio in the summer of 1960, when 200,000 children received the vaccine orally, on sugar lumps or in syrup, on what became known as 'Sabin Sundays'. An experienced insider, Sabin lobbied hard for his vaccine, arguing that an oral vaccine would lead to greater take-up rates, as it was easier to administer than Salk's three injections. Sabin's vaccine had another crucial advantage: it passed through faeces to provide a kind of immunity to those in the community who encountered it. After an investigation, and under pressure from drugs companies who favoured Sabin's new product, the American Medical Association recommended the Salk vaccine be replaced. The Sabin vaccine was licensed as its replacement in September 1961.

Despite claiming that their differences were based on purely scientific disagreements about the effectiveness of a killed vaccine over a live vaccine, there was often an unseemly bitterness and envy at the heart of Sabin's attacks on Salk. 'Salk was a kitchen chemist,' Sabin sniped nastily: 'He never had an original idea in his life. . . You could go into the kitchen and do what he did.'[29]

This sentiment was widely shared in the scientific community, which resolutely refused to honour Salk's achievements; even declining to invite him to join the National Academy of Sciences. His colleagues opposed the principle of his inactivated vaccine, and they opposed his status as a media darling even more. One scientist present at the announcement of the Salk vaccine in Ann Arbor recalled: 'The bedlam was disgusting . . . It was as if four supermarkets were having their premieres on the same day.' Sour words for a day that saved millions of people from death and paralysis. (By contrast microbiologist Maurice Hilleman was called the 'forgotten pioneer of vaccines' by the *New York Times* in 2013 after scientists claimed he had saved more lives in the 20th Century than any other individual. Hilleman, a modest man who said his proudest achievement was 'being able to survive while being a bastard', developed over forty vaccines during his lifetime, including ones that remain in the vaccine schedule today for measles, mumps, hepatitis A, hepatitis B, chickenpox and meningitis.)

By the mid-1960s Jonas Salk had retreated to the newly created Salk Institute near San Diego. Eventually he divorced his wife Donna to marry Francois Gilot, an artist who had borne three children by Pablo Picasso. He gave up his white lab coat, grew his hair and took to sporting turtlenecks. Much to his dismay Salk never lived to see his pioneering vaccine again in widespread use. 'Normally my father tried to let these things go,' Salk's son Peter remembered, 'but this one was so terribly painful, so personally insulting to him as a scientist, that he couldn't let go. It is not exaggeration to say that it haunted him for the rest of his life.'[30]

Salk could not have foreseen that his rivalry with Sabin would continue after both of them were dead. Although Sabin fought the conclusion for the rest of his career, there was undeniable evidence that his oral polio vaccine caused a handful of polio cases in those who took it or encountered it. Between 1965 and 1985 the USA recorded eight to eighteen cases of 'vaccine-associated paralytic polio' (VAPP) where the attenuated virus had regained some of its virulence. Several hundreds of VAPP cases have appeared elsewhere in the world, with some more worrying cases of hybrid viruses, which recombined the OPV strain and a wild strain, and spread rapidly. Sabin's vaccine was withdrawn in the USA in 1979. As the fight against polio turned into a global eradication campaign, only Salk's inactivated vaccine could achieve the final steps in wiping out the disease.

Children around the world have consumed more than 10 billion drops of oral polio vaccine since the global campaign began in 1988; approximately 12 doses a second.

3

POLIO'S LAST STAND

The polio virus was an unseen agent of terror during the first half of the twentieth century; killing some, paralysing many more and horribly condemning others to 'iron lungs' where they sometimes spent years unable even to take a sip of water by themselves. At its height, eight thousand children a year were paralysed in the UK, but with the development of the Salk Inactivated Polio Vaccine (IPV) and the Sabin oral polio vaccine (OPV), it appeared that polio was a disease of the past. Other countries, such as Communist Cuba, conducted extensive national campaigns, using OPV to vaccinate every child repeatedly over several years, and proving that the disease could be entirely eliminated. Similar vaccination campaigns in Central and South America demonstrated equally spectacular results.

'Rotary – go home. . .'
Who decided to rid the world of polio? Not politicians or global health organizations, as you might expect. The starting gun was fired by Rotary International, a network of businessmen more used

to enjoying convivial dinners, raising money for local good causes and organizing floats to carry Santa Claus around suburban neighbourhoods at Christmas.

One Rotarian, Dr John Sevrer, ran a laboratory and clinical research programme at the National Institute for Health in Bethesda, Maryland, studying viral infections in children. He discussed the success of the vaccination programme in Cuba with Albert Sabin: 'Dr Sabin had been to Cuba in the mid-1960s and was very impressed with the idea of immunizing everybody,' he recalled.[1] Under Fidel Castro's leadership, in early 1962 Cuba introduced National Immunization Days, organizing vaccinators into 82,000 local committees that administered the vaccine to all children on the island twice a year, every year. The results were startling: no cases of wild poliovirus were reported after 1963.

Persuaded by Sabin's comments and his own research, Sevrer suggested that Rotary adopt polio immunization as its fundraising mission:

> It was obvious that we could have a major impact bringing immunization to the people of the world. About half of the children in the world were not receiving any vaccines and polio vaccine was the vaccine I recommended that we concentrated on. It could be given easily by just two drops in the mouth of the child, it was relatively inexpensive at about four cents a dose and we had Rotarians throughout the world who could all help by directly participating in the immunization programme, physically by their presence and

> through their contributions. After all, you
> could train somebody to be a vaccinator if
> they could count to two, because all they'd
> have to do is put two drops in the mouth.[2]

Rotary International, seeking an ambitious programme to mark the organization's seventy-fifth anniversary, decided after much discussion to support a campaign to target Health, Hunger and Humanity (3-H). The main proponent of the initially sketchy scheme to 'do something lasting in the world' (as a committee member, Clifford L. Dochterman, put it) was that year's Rotary's international president, Clem Renouf. Inspired by the smallpox campaign, Renouf asked John Sevrer what other disease might be a candidate for eradication. Sevrer replied that 'if a single vaccine were to be selected for the 3-H programme, I would recommend polio-myelitis'. Renouf proposed the campaign to the 1979 board meeting and the organization agreed that polio immunization would be its international mission; a mission that remains to this day.

The campaign began in Asia. 'Ben Santos was a Rotarian and an ophthalmologist from the Philippines and he volunteered that might be a good place to start,' Sevrer said. Dr Sabino (Ben) Santos had good government contacts and could organize the approvals and cooperation needed for a major immunization campaign. In addition, the Philippines had the highest incidence of polio in the Western Pacific but its new national immunization programme did not include the polio vaccine. Rotary agreed to provide about US$750,000 worth of vaccines and pledged that Rotarians would also participate in the immunization days to get the vaccine out to the children.

Not everyone was convinced that Rotary was capable of the task but proof came from its organization of a mass immunization campaign against maternal tetanus. The polio programme was approved and the campaign got under way. On 29 September 1979 Rotary's incoming president, James L. Bomar, vaccinated the first child against polio in Makati, Manila. The little girl squirmed so much in her mother's arms that Bomar could hardly squeeze in the drops. When the job was done, a group of polio-stricken boys playing in the mud nearby called out 'thank you!'

Six and half million children were vaccinated in Rotary's first international polio project, which took place within the larger Expanded Programme on Immunization (EPI). Within the first two years of the campaign the number of polio cases decreased by sixty-eight per cent. Just as in the 1940s and 1950s' March of Dimes, the global fight against polio was once again led by people who initially knew little about the disease but could marshal widespread popular support.

However, although Rotarians believed a polio vaccination campaign was a good idea, their enthusiasm was not shared by the global health community, where a deep sense of gloom over the failure to wipe out malaria had kept single-disease eradication off the agenda. In addition, Rotary proposed taking a completely different course to that set out by health leaders, who had rejected costly eradication campaigns in favour of building local health services. John Sevrer said:

> The WHO was in favour of immunization but not specific to any one disease and it was going step by step across various countries. We were coming in with a vertical programme, working specifically on polio and then bringing in other vaccines after it.

In a clash of competing visions, WHO director general Halfdan Mahler, believed he knew best. Mahler shifted the work of the agency away from supporting expensive hospitals in large cities towards basic health services accessible to everyone. Despite the availability of cheap vaccines for diseases including diphtheria, whooping cough, tetanus, measles, tuberculosis and polio, in the 1970s fewer than five per cent of children in the developing world were fully immunized. Countries either lacked the foreign currency needed to buy vaccines, or could not transport and refrigerate them properly. In this context, Mahler saw routine immunization as a good way of building basic health infrastructure, by encouraging women to regularly visit local clinics with their babies. But he believed neither in the principle of single disease eradication campaigns nor in Rotary's ability to stay the course and carry one out.

At a meeting before the World Health Assembly in Geneva in 1984, Mahler bluntly told Rotary:

> Volunteer organizations have approached us
> like this before. Do-gooders make promises
> but fail to follow through. We have seen it so
> often, a volunteer group comes to a country,
> starts a programme and then leaves. So,
> Rotary, we appreciate your interest but we
> would like you to go home.[3]

With the future of the polio programme in doubt, Albert Sabin was co-opted to make the case. Unfortunately, when Sabin met Rafe Henderson, the WHO's director of the Expanded Programme on Immunization, in Washington DC, they immediately clashed over the concept of pursuing another single-disease campaign. Sabin slammed his cane on the table, raged that Henderson had always opposed his ideas, and stormed out.

No one denied that, with an effective vaccine on the market and no animal reservoir, polio was a good target for eradication but Henderson shared the qualms of many others; scientists remained divided about whether eradication of a disease in which ninety-five per cent of the victims were 'invisible carriers', with no symptoms, was really possible. And the vaccine itself could be unreliable, losing its potency quickly unless kept refrigerated at the correct temperature, something that made it difficult to distribute in some parts of the world.

The global public health community did not agree that either country-wide elimination or global eradication of polio were realistic targets but the truth was that Rotary's campaign worked on the ground. Successful campaigns, first in the Philippines and then in Central and South America, opened the door to the idea that eradication was possible. 'People wanted their children protected against polio,' John Sevrer said:

> The advertising would be on the radio or in the newspapers but a lot of people didn't read. We had to employ town criers walking through the streets telling parents about it. In other places they would have a big parade in the middle of the village and tell people about it that way.

The polio eradication campaign in the Americas was led by Dr Ciro A. de Quadros, who led the EPI programme in the region for the Pan-American Health Organization (WHO's regional office). De Quadros had successfully eradicated smallpox from Parana state in Brazil and in Ethiopia and was in receipt of Rotary grants for routine immunization. Between 1978 and 1984 the number of children in the Western Hemisphere paralysed by polio fell by ninety per cent. In Brazil (a difficult country due to its vast, challenging terrain), they fell from 2,330 per year in the late 1970s to only 45 in 1983 (an even more staggering 98%). Crucially, de Quadros's work not only demonstrated that national polio campaigns were successful but also that they could contribute to strengthening basic health systems. Rotary could return to the WHO with evidence from Ciro de Quadros that polio immunization had increased vaccination against other diseases and increased national planning, political will and funding.

By the late 1980s even the naysayers had to acknowledge that Rotary had been more successful than it could have imagined. The most contentious word of all – eradication – was within the realm of possibility. In 1982, at the suggestion of John Sevrer, Rotary took another step, adopting the goal of immunizing all the world's children against polio by 2005, followed in 1985 by a pledge to fund a global polio immunization campaign that ultimately raised US$250 million.

A decade after Rotary seized the baton for polio immunization, the world heath community finally changed its mind. On 29 May 1988, the World Health Assembly considered the evidence of the successful campaign in the Americas and voted to eradicate polio by 2000. At a meeting in Talloires in France observers noted, 'A light seemed to go on in Halfdan Mahler's brain.' The head of the WHO 'undertook a 180 degree switch in his thinking', producing a piece of chalk to draw a diagram of how he believed the polio campaign could build health infrastructure.[4] A campaign organized by ordinary men and women had taken on global bureaucracy and won.

'Nothing is beyond you, even in India . . .'

The polio eradication campaign brought together four partners – UNICEF, WHO, Rotary International and the CDC – as the Global Polio Eradication Initiative (GPEI) to cooperate in tackling the disease. Headed by a veteran of the smallpox campaign, Nick Ward, and Canadian epidemiologist Bruce Aylward who led the GPEI from 1998–2002, the WHO began constructing a surveillance network, including reporting, investigation and a laboratory network that, by the 1990s, included five specialized laboratories, fifteen regional laboratories and sixty national laboratories spread over many countries. The WHO's new plan for polio eradication, drawn up by a group chaired by the initially sceptical D.A. Henderson, outlined how activities would be organized by region, with every country immunizing ninety per cent of its children against polio by 2000. Polio eradication zones would then be established and enlarged.

Success was dramatic. The number of polio cases fell from 350,000 a year in 125 countries, to only 400 cases by 2013. 'Over the first years there was a very rapid decrease in the number of cases of polio worldwide and it was very important to get that going,' John Sevrer said. 'The big drop was about the years 2000 to 2002. Since then we've had small drops but we're not down to zero and zero is the target.'[5]

The initial gains were huge but, even with international backing, the campaign progressed only in fits and starts. Everything and anything could throw it off course: competing government demands, war, economic crises or the logistical challenges of co-ordinating a global campaign, involving many organizations, in countries with vastly different levels of healthcare.

Chris Maher is a veteran epidemiologist and WHO senior adviser who weathered many of the storms of the polio campaign. 'When we started doing eradication we started off in countries where the infrastructure was not too bad and where there was a reasonable tradition of delivering public services,' Maher said:

> The Americas were the first to eradicate, next were countries like China and Vietnam and Malaysia; big countries with a lot of challenges but fundamentally reasonable systems for delivering services. Smaller countries like Papua New Guinea, Laos and Cambodia were more difficult, not impossible. That was the easy phase of eradication.[6]

The first countries fell like dominoes but the sense of rapid progress soon evaporated. Next came India, the country where wiping out smallpox had almost proved impossible. 'When the campaign focused on South Asia, the scale of the problem became quite terrifying,' Maher said. Sub-national rounds in India targeted somewhere in the region of fifty million children, rising to one hundred million in national rounds. In addition, the speed of population growth was 'pumping kids into the environment' faster than healthcare workers could reach them. International healthcare workers, such as Maher, soon saw that the proper working of the immunization programme was hampered by huge geographical and staffing problems.

'India was the most polio endemic country for thousands of years,' said the Pakistan-born doctor, Hamid Jafari, who ran the campaign in India for six years and led the polio programme at the WHO until January 2016. 'In that context, because of its size and because of its importance in the world, it was also a very important source of international spread of polio.' The country struggled under a burden of up to 150,000 polio cases at a time. Its dense population, poor sanitation and climate created ideal conditions for the disease to thrive. Jafari admitted: 'It took an extraordinary effort to achieve eradication in India.'[7]

A stop-off by a UNICEF polio eradication team on a sleepy afternoon in the Agra suburb of Sarai Khwaja in December 2011 demonstrated both the complexity of the campaign and the strides needed to reach its target.

In a Muslim district of the city, a man stood on a stepladder, fixing an unfeasibly tangled bunch of electric wires with a pair of pliers. A monkey eyed him from a nearby windowsill. Women sat on rope daybeds, chatting. A small team of polio coordinators mingled, talking to people and showing an educational film about preventing the disease. A group of mothers and their small children eagerly gathered at one house to watch the film; they'd heard the message before but it's good entertainment. A gutter brimming with filthy water and sewage ran directly in front of the door, near where the children played; an everyday and common breeding ground for polio, once very common in this community. A father said that, like many Muslims, he was reluctant to accept the polio vaccine at first, believing that it would make his children sterile. But the intervention of the local Imam on behalf of the polio co-ordinators changed his mind; now he is happy to have his children vaccinated.

Thanks to such grassroots effort, polio is a much rarer occurrence in the area but only one street away, a victim of the disease was confined to her parents' home. Sitting in the main room of her house, seventeen-year-old Saima is an example of a life thwarted. In the dim room, where the only light is reflected off a wall of metal platters, Saima brightly conducted a conversation from her rope bed. Whatever hopes her family had for her future are over. She will never marry or earn a living: a disaster in her community.

The serious physical consequences caused by polio in Europe and the USA are magnified in countries such as India, where it wrecks lives and families. 'A whole life alters after you get polio,' said Hamid Jafari. He remembered treating his first polio patients in Punjab province in his native Pakistan:

> Young mothers were coming to me with young infants who were paralysed and those mothers had all of this hope in their eyes that maybe polio could be cured. They wanted me to treat their babies but I knew they were looking at lifelong paralysis. That's heartbreaking and that's part of the inspiration and motivation that keeps you going.[8]

India's slow progress was infuriating. The journey of polio eradication in India had started in 1978, when oral polio vaccine became available, but took more than thirty years to succeed. A paediatrician, Naveen Thacker, said: 'In 1992 I still saw five cases in a month and it drove me mad.'[9] A decade after the start of the eradication campaign, two areas of the country still sustained transmission: the western part of Uttar Pradesh and the central part of Bihar, home to more than thirty million children.

Success or failure hinged on finding a way of reaching those children. It would take a complicated partnership of international and local non-governmental organizations (NGOs) working with the Indian government to employ an array of techniques, including an unknown level of micro-planning and satellite mapping to ensure that every house in even the most far-flung community was receiving the vaccine, deploying mobile teams of vaccinators to track transient communities, using finger markings, and working at bus stops or at sites such as brick kilns where migrant workers and small children could usually be found.

'There was a whole string of innovations that were brought in to try to improve the quality of the immunization campaign,'[10] Chris Maher reported. Healthcare workers could no longer rely on parents to tell them the truth about whether their children had been immunized. In some areas there was a general lack of community engagement, in others an outright hostility from minority groups towards the Indian government and all its endeavours.

House marking was used as a tactic to cover very large communities. This dramatically improved the monitoring and delivery of the campaign. 'The immunization team would mark a house with a piece of chalk, designating which campaign it was, which direction they were going in, which house number this was and then micro plan,' Maher said.[11] Health workers went house to house delivering the vaccine rather than setting up fixed immunization points and waiting for people to come to them. Children who had been vaccinated had their fingers marked with indelible ink.

The campaign in India was also at the forefront of developing disease surveillance. Naveen Thacker said:

> We were able to use technology for tracking the virus by molecular biology with state-of-the-art laboratories in Mumbai, like the Enterovirus Research Centre. We did a lot of operational research and serosurveys [drawing blood from children] so that we knew the origin of the problem.

In addition to sereotyping, the campaign deployed environmental surveillance, which had first been introduced in Finland and Israel in the 1980s. 'We started tracking wild virus not only in children but also through sewer samples. In that way you can detect the virus in sewers before you get a case in a person,' Thacker said.[12]

More than two million volunteers were mobilized for each national immunization day, setting up 640,000 vaccination booths and vaccinating approximately 172 million children on each occasion. Thacker said:

> In the process of the polio campaign we built up a huge network of surveillance medical officers in the form of the National Polio Surveillance Project, a social network through the media and a large number of volunteers posted in Uttar Pradesh and Bihar, where they worked for polio for many years to mobilize people.[13]

Driving back through Agra after the afternoon's efforts in Sarai Khwaja, one health worker from an international NGO described the effort to eradicate the disease as the equivalent of building a mammoth 'shadow' national health system of more than 900,000 health workers. India's per capita healthcare spending was among the lowest in the world, at only $43, compared to $87 in Sri Lanka and $155 in China.

Chris Maher clarified that he believes it was not so much the establishment of a parallel service but more the case of adding another layer on top of it:

> Even in places where the system was not functioning very well, the basic delivery was still based on the system. It was still largely run by the structures of the government. It's

just that those structures were not at a level
necessary to eradicate polio.[14]

'The onus was on the government of India,' according to Naveen
Thacker. 'International agencies provided their expertise but the
campaign was led and driven by the Indian health system and the
Ministry of Health and Family Welfare. There was the highest
level of political commitment from top to bottom in India.'[15] That
commitment faltered at times, however, just as it had done in the
smallpox campaign. Maher said: 'There were some terribly hard
times in India when sometimes we weren't sure whether the government was going to keep going or whether anyone could keep
going. It was a question of determination and just hanging on.'

Polio was tenacious. If the quality of the campaign or surveillance dropped for a few months, the disease returned with a
vengeance. 'You'd be punished for it,' He continued:

> It's a very stressful thing to be doing. We were
> putting in a massive effort and spending a
> huge amount of money. Every few weeks
> we'd send out hundreds of thousands of
> vaccinators and immunize tens of millions
> of kids and then we'd get to the high season
> and still find polio transmission.

There appeared to be no relationship between effort and progress.
Even in a good year, a fresh outbreak could pop up. The years
between 2005 and 2007 were a rough period, Maher admitted, as
was 2009 to 2011. Fresh outbreaks sprang up, depressing everyone.
'That's one of the things that really takes the wind out of the sails
of partners and funders,' Maher said:

People on the ground were less depressed because they realized that this happens. But it's very disturbing when you're a little bit removed from it and you see slippage. Those two periods created a lot of turmoil in the global polio programme and among donors. The turmoil was related to doubt about whether or not we were doing the right thing, were we going about it the right way and were we actually going to succeed in eradicating.[16]

On the ground the campaign seemed bogged down in years of effort and toil for little benefit. Sumita Thapar worked as a media consultant for UNICEF on the polio programme from 2009 to 2012 and found communicating the same messages a day to day struggle, especially compared to her previous work on HIV/AIDS:

From a development perspective the polio programme was seen as being very boring. The programme had been going on for years and years, while AIDS was something where there was a lot happening and the discussions were moving forward very fast. With polio one felt that things hadn't moved forward. If you looked at the advertising campaigns over the years they were saying the same things over and over again.

India's HIV/AIDS programme was inclusive and focused on human rights and dignity, something she believed clashed very much with what she saw in the polio campaign:

> In polio there was a top-down approach in
> the field. This was something that we were
> just imposing on people. People would look at
> us and say, 'Oh you've come again but you
> came just last month and we don't have clean
> drinking water, we don't have toilets, we
> don't have this, we don't have that; *and all you
> can think about is polio.*[17]

If parents did not want to immunize their child the mayor of the town would be summoned to pressure them. 'Not exactly forced but pressurized', Thapar said. When she shared her disquiet with the lack of what she believed was real community engagement, she encountered strong opposition from her colleagues: 'They would disagree with me. They were really angry and they said that there were many community leaders and religious leaders and education institutions involved. But I didn't see the same passion or the same energy.'

In India, Thapar believes polio was stigmatized as a poor person's disease that afflicted the poorest communities. 'There was a class thing. With AIDS, everyone was interested because everyone was vulnerable. But polio was really seen as an issue that affected only the poor and marginalized.' To make matters worse, when Thapar encountered polio survivors, she had little to offer them, even if they had contracted a vaccine-related form of the disease, caught from the OPV vaccine: 'The government should have taken responsibility for that and offered them something.' Most jarring, she said, was the lack of respect shown to polio sufferers:

> Every time they held a rally ahead of a polio
> round they would put two polio-affected

people, sometimes on a single wheelchair, with garlands around them and they would lead the rally. They were obviously poor people and the message seemed to be that if you don't immunize your child this is who your child will become.

The people who attended those rallies, Thapar said, were health workers or schoolchildren, obeying orders. 'They didn't have a choice. They were told to come, they came. It wasn't as though people really believed that polio must be eradicated. I didn't see that at all. It was just people who were there because they needed to be there.'

Among everyday workers, like Sumita Thapar, there was no doubt that fatigue and frustration was growing. Would they ever be able to eliminate the disease? Even Bangladesh had managed it but somehow India could not. As the campaign eventually drew towards a successful conclusion the feeling was one of relief, rather than jubilation. 'Even in India, with so many problems, nothing is beyond you,' said the coordinator of the polio campaign from the Indian government in New Delhi.[18] Effective use of surveillance data and emergency planning was critical, as was the personal flattery of local officials and knowing how to grease the wheels of local bureaucracy. People needed to know who to call, what to threaten, and how much to offer as an incentive.

The campaign steamrolled forwards. When a polio outbreak threatened to engulf Calcutta, a swift response vaccinated 100,000 children in six days. When a World Bank loan threatened complicated funding arrangements for the OPV the Indian government was persuaded to turn down the loan to avoid jeopardizing the campaign. The momentum was unstoppable and final victory sweet. In April 2014, India was declared polio-free, three years after reporting the last case.

The announcement of a polio-free India and, as a result, a polio-free South-East Asia, was a halcyon day for the Global Polio Eradication Initiative (GPEI). But while the overall campaign seemed momentarily stunned and dazed by the success, workers on the ground were wrung out and exhausted after intense years of battling seemingly intractable problems. India had taken its toll. 'One of the things that complicated the global eradication effort from that point on was the level of fatigue,' Chris Maher said:

> There was a level of fatigue among donors, in governments, among the partnerships who'd been working on polio and a level of impatience that things were taking longer than everyone wanted. We just needed to get the thing finished.

Endemic polio remained in only three countries (Afghanistan, Pakistan and Nigeria); the final stages of global eradication seemed within reach but also hopelessly far away. 'There was a period of reinvention when it became obvious that India was going to close things out,' Maher said. 'The global programme tried to change the way it looked at things and do things differently, because Nigeria and Pakistan were very different challenges.'[19] The breakthrough tactics developed in the campaign in India: engaging communities, dealing with minority groups and implementing micro-planning, could be used in the global eradication campaign, transplanted successfully to other countries and used for other diseases, including measles, diarrhoea and Ebola. Victory in India had shown what was possible. There were no excuses left: 'Once the programme in India was able to stop the transmission of polio virus, there was no longer a biological reason or a question of feasibility, that polio could be eradicated,' said Hamid Jafari. 'The challenges then were social and political.'[20]

'Like an arranged marriage, where the love comes later'

India was tough but what came next tested Jafari's 'no excuses' theory to the limit. The programme had to tackle polio in one of the most war-torn and inhospitable countries in the world: Afghanistan. It could only do so by implementing lessons learned the hard way in India.

At the helm was Peter Crowley, who headed the UNICEF programme in Afghanistan for more than three years. No sooner had he taken up the job when a huge increase in cases in 2011 gave the polio campaign a very bad year; he had to radically rethink the structure of the programme and galvanize the team into action.

'Afghanistan is quite an extraordinary place in terms of terrain,' Crowley said. 'When you fly over it, it's predominantly brown, apart from areas of cultivation or river valleys. A lot of it is desert or semi-desert but then you also have these extraordinary high mountains, the Hindu Kush mountains, that run through the middle of the country.'[21] In the harsh winters of the mountains, teams faced the physical challenges of getting from place to place by foot, bicycle and motorbike. Crowley found, however, that difficult terrain was the least of his problems. The high mountains or harsh winters didn't stop the vaccinators:

> It was more to do with conflict and insecurity
> and the fact that just moving from place to
> place is complicated and dangerous and
> requires people to be confident that, for

example, the Taliban or the international military know that there is a campaign going on and that the teams are going to be moving around.

And sometimes, Crowley discovered, the problem wasn't even security.

The main centre of transmission in Afghanistan was in the south of the country, in Kandahar, Helmand, Urozgan and Zabul, where the conflict was at its height. Crowley discovered that teams sometimes reported areas as inaccessible when the truth was they just didn't want to go there. 'It was simply an operational failure. They wanted their pay but they didn't want to do the work and thought they could get away with it by claiming insecurity.'

Crowley restructured the office in the south, bringing in new staff who worked together with staff from WHO, side by side, in a central control room; an obvious but, in practice, revolutionary concept, Crowley admitted:

> We're both UN agencies so in principle we're both guided by the same charter and the same values and the same principles; we work alongside each other all over the world. Nevertheless, we are different organizations with different cultures, so it's not always easy to be on the same wavelength on all issues.

Sharing the same office space and working as a combined team in Afghanistan was far from easy at the beginning. 'It was like an arranged marriage in which the love comes later,' Crowley said 'but the relationship has got a lot better.'

Within the GPEI partnership the WHO took responsibility for surveillance, managing the teams of vaccinators, leading the process of micro-planning of campaigns and providing the technical health components of the programme; UNICEF ran social mobilization and communications and organized the buying and transport of the vaccine.

'Until I joined the polio team full time I thought it was a bit like Amazon,' Crowley said, 'You place your order and the vaccine turns up!' In fact, the process is enormously complicated, with a limited number of global suppliers producing a finite supply every year. Orders must be planned a couple of years in advance, taking into consideration the constantly changing picture of the epidemiology of the virus, or a sudden multi-country outbreak.

In terms of communication, Crowley said, 'We obviously need to understand how the communities perceive polio and how they prioritize polio vaccination for their children and the extent that there may be reservations or objections.' In Afghanistan there was less resistance to polio vaccination than in other countries, with the main challenge lying in reaching the children: 'The big problem in Afghanistan is that we can reach the household but not necessarily reach the children of that household. Their parents say they're sick or they're sleeping or they're too young to have the vaccine.' With men not allowed to interact with Pashtun women, recruiting mixed-sex vaccination teams was important but difficult. Afghan families did not want women to be out 'wandering about in the community,' Crowley said. Although progress has been made, Crowley said the situation is still problematic: 'We wanted to halt transmission so we had to reach every child. We needed to find smart ways of getting around those barriers.'

Some areas were plagued by insecurity, with on-going fighting;

in other areas it was difficult to establish who the community leaders were:

> The Taliban in Afghanistan were generally supportive of the eradication effort but they don't have very clear lines of command and control, so you could have a situation where a local commander for their own reasons was opposing the campaign. Or insecurity was caused by criminality or something totally separate from the anti-government insurgency . . . There was an awful lot of work done in just trying to understand who the local leadership was, how to reach them and then what their issues were but this proved to be very successful.

Quite suddenly, the campaign was reaching areas that hadn't been touched for four or five years and cutting down dramatically on the number of missed children. The polio team wanted the local people to feel they were in charge. 'When I first went down to Kandahar in early 2010 I met the governor and other local officials and polio was not high on their agendas,' Crowley said. 'But when I made my last visit to Kandahar and Helmand in 2013 it was quite extraordinary to see the different level of commitment from the governor, the police, the security forces, the district assembly, religious leaders and teachers; everybody.' Local people wanted to end polio themselves, with men from some of the most insecure areas turning up and asking for vaccine to take back to their villages. In Afghanistan, the polio campaign earned the trust of the people; elsewhere, lack of trust would be its downfall.

The programme works, Crowley says, 'as long as they know that there's no ulterior motive, that we genuinely have the best interests of their children at heart, that they can trust the vaccinators who come to their doorsteps and that they can trust that the vaccine is good for the children.' Over the border in Pakistan, a dramatic turn of events was to destroy the trust polio workers had spent so long building and endanger the entire global polio eradication effort.

'Pakistan: polio central'

Since the development of Albert Sabin's 'Red' vaccine, tested in the USSR, and the success of early eradication campaigns in Communist states such as Cuba, polio has been a curiously 'political' disease. Those political stakes were dramatically heightened after it was discovered that the CIA, in an effort to locate Osama bin Laden, had, with the help of a Pakistani doctor, Shakil Afridi, instigated a fake door-to-door hepatitis B vaccination campaign. The plot aimed to obtain DNA from bin Laden's children to prove the family was living in a local compound. Afridi was arrested by Pakistan's Inter-Services Intelligence agency (ISI) for working with American intelligence agents and sentenced to twenty-three years in jail. The situation worsened when the Kathryn Bigelow film, *Zero Dark Thirty*, wrongly identified Afridi as a polio vaccinator. In response, factions of the Pakistan Taliban decided to ban vaccination in the north-west tribal areas of Waziristan and the Khyber Pakhtunkhwa province.

'Pakistan is polio central right now,' said Chris Maher, after the ban quickly led to a tenfold increase in polio cases (which had fallen to six the previous year) and the shocking murders of more than sixty polio workers in the two years between 2012 and 2014. Claims that the polio campaign gathers information and spies for the US are used by rebels in other parts of the world to justify attacks on polio workers. 'Polio has become a very, very political issue, rather than a humanitarian one,' said Sir Liam Donaldson, who chairs the international monitoring board of the Global Polio Eradication Initiative (GPEI). 'We need to move it back along the spectrum to becoming humanitarian again.'[22]

There was a sense of relief in the global health community in May 2014 when President Obama's senior counter-terrorism and homeland security adviser, Lisa Monaco, replied to the unhappy leaders of twelve prominent public health schools confirming that the CIA would not use immunization programmes or workers as a means to collect intelligence. 'The director of the Central Intelligence Agency directed in August 2013 that the agency make no operational use of vaccination programmes, which includes vaccination workers,' Monaco wrote. 'Similarly, the agency will not seek to obtain or exploit DNA or other genetic material acquired through such programmes. This CIA policy applies worldwide and to US and non-US persons alike.'[23]

Speaking in her office at the CDC in Atlanta on the day Monaco's statement was released, the Global Immunization director, Rebecca Martin, was relieved, but annoyed the situation had ever arisen. 'We see it as a very positive statement,' she said, adding that it offered some 'reassurance of what could be done in terms of ensuring the safety of vaccinators and making sure that the work of intelligence is not related to public health interventions.' It should never have happened but there was hope, she said, that the Taliban ban would be lifted. 'We've seen some of the discussions between the military and the civil government in Pakistan about how to start looking at that ban. They are vaccinating at transfer points around the banned area, so that's a start. But there's more that needs to be done.'[24]

If negotiations were going on behind the scenes, Peter Crowley admitted no one from the polio programme could exert much influence to lift the ban itself:

> I doubt it, frankly. There are quiet efforts going on, including in the margins of the peace talks that are taking place between the government of Pakistan and the anti-government element in these areas. The symbolic importance of the ban is too great to those who have the biggest stake in it. They're not going to easily let it go . . . At the same time the people who control this territory are themselves the parents of children.[25]

While the ban remained in place the polio campaign could only work around its limits, vaccinating at transfer points or smuggling in vaccine when local people requested it. 'These strategies will reach some children,' Rebecca Martin said, 'but the ban needs to be lifted to reach all the children. We hear from our colleagues on the ground that families want vaccine, they see their children being paralysed and it doesn't need to be that way.'[26]

Throughout the summer of 2014 the polio team worked hard to come up with ways of beating the ban, such as using a breakthrough in the security situation in the Khyber Pakhtunkhwa province to swoop in and vaccinate seven million children within three months. But while the ban remained, transmission could not be stopped. 'I think the entire country is under threat from the areas where there is uncontrolled transmission in the federally administered tribal areas of Pakistan,' a rather weary Hamid Jafari said during a stopover in London. 'There's a polio outbreak going on but no vaccination happening at the same time.'[27] Not only was the situation imperilling the polio programme in Pakistan but the global campaign was also under threat as polio strains spread across the globe to the Middle East and Africa.

'We should have been done with polio about one-and-a-half years ago,' the businessman and local Rotary International chair Aziz Memon admitted down a crackly phone line from Karachi. In five and half years he saw Pakistan move close to the brink of erad-icating polio, only for all the gains to be reversed by the ban and the series of brutal killings. 'Unfortunately on 18 December 2012 there were some killings of vaccinators in Karachi and in Peshawar. This distracted the campaign and created a negative impact on the other vaccinators.'[28]

Memon said that one family in Karachi lost two daughters, murdered as they volunteered as polio vaccinators. Their younger sister now goes from house to house, giving polio drops herself, telling people that this was the mission of her two sisters. 'This is the sort of spirit which makes us great,' Memon said. 'Yes, it is dangerous but they go.'

The willingness of polio vaccinators to risk their lives has become the most extraordinary element of the eradication cam-paign. It is something no one could, or would, have wanted to predict. 'It really is quite amazing,' Peter Crowley said. 'I was inter-viewing somebody for a job recently and she said she'd read about the story of women going door to door in southern Afghanistan and was so inspired by their bravery and commitment that she felt she wanted to be part of the effort.'[29]

The murder of a further two polio workers in Pakistan in January 2014 prompted new tactics. Instead of just guarding the vaccinators, entire areas were cordoned off and the motorbikes often used in attacks were banned. But tweaking security wasn't enough. With the number of cases rising, the GPEI began a complete overhaul of the polio campaign in Pakistan. Rotary buckled down, supplying 135,000 vaccine carriers, setting up cell phone monitoring, building nine new centres for community awareness in high risk areas, setting up water filtration plants and providing routine immunization and general health camps in border areas where there were large populations of migrants.

Crucially the campaign also negotiated for the protection and assistance of the army, with army medics delivering the vaccine in areas where it was not safe for polio workers to go. 'We are in touch with the Taliban,' Memon said. 'Officially they are saying that they have not issued any ban against the polio campaign and are not killing the vaccinators. But the Taliban has several splinter groups and the main Taliban is not taking responsibility for the killing of all the vaccinators.'

The major challenge remained finding a way to give vaccine to the children trapped in areas affected by the ban. Memon confirmed the harsh reality: 'Nearly all of the cases are Pashto and have originated from an area where there has not been a single round of vaccinations for more than three years.'[30] To break through, the polio programme in Pakistan needed to go 'back to basics' in the summer of 2014. One year later, the country was making progress.

At a staff meeting in a grey and silent conference room in the Gates Foundation offices in Seattle, the deputy director of the polio team, Tim Petersen, outlined the ongoing areas of transmission in the Khyber corridor and charted the fall and then rise of polio cases. 'Things were getting better pre-2012,' he said 'and then there's this time right here in December where things changed.' He points to one spot: 'It's where the attacks on the vaccinators occurred. We had the ban then in North and South Waziristan. Then we had a change in leadership of the government a little bit later here in the year.' Another point on the chart shows the number of cases rising dramatically but then falling. Later, Petersen said: 'There are a lot of things that happened in the programme that made it more difficult to be able to deliver vaccines to all children. It's taken a while to move past that.'

A change in political leadership brought a new prime minister, who took some time to get up to speed, but stronger government leadership and a prime ministerial focus group resulted in the sharing of information between the Ministry of Defence and the Ministry of Health. National emergency operation centres, and provincial operation centres in twelve high-risk priority districts, were set up. The polio campaign benefited from an army campaign to push terrorists out of North and South Waziristan that allowed vaccinators to start working again. When the political machinery works, operational changes follow, Petersen said: 'Micro-plans hadn't been updated for years because of the security situation when we weren't able to get into the field. People come and go, target populations changed and high-risk groups within certain areas changed.'[31]

Like India, Pakistan had brick kilns, construction sites and large numbers of people who migrated for work; people who had not been captured on micro-plans for a long time. This moving population is one that vaccination campaigns or routine immunization often miss. Hundreds of vaccination posts were set up at major transit routes, railway stations and bus stops, vaccinating more than nineteen million people who were on the move. All the teams of front-line health workers have been retrained and the programme is trying to redefine how polio is presented to the community: 'A vaccinator is presented as a community protector: somebody who's there to help,' Petersen said.[32] Women from the local communities in Karachi, Central Sindh, Balochistan and the tribal areas are now working as 'community protected vaccinators'. 'We did it before, we just didn't do a very good job of doing it,' Petersen admitted:

> A lot of times team selection wasn't always as it should be. Sometimes they were hiring relatives or sometimes hiring vaccinators from outside because it was easier to send health workers from a clinic than actually go and sit down with the leaders in that community and identify the right people. It requires more time and effort to do that.

Mobile phones were used to send real-time data before and during campaigns and provide instant reports on laboratory samples. In addition, the campaign agreed to cut back the number of vaccine rounds to nine a year, allowing a little more breathing space to fix problems when they arose. The polio campaign gradually peeled away the layers of the onion, digging deeper into the drivers of transmission in the remaining areas. 'We're going into Peshawar and we're finding Afghan communities that haven't ever been vaccinated or included in the micro-plans,' Petersen said:

> There was an Afghan family who had come to Peshawar for a wedding. They had left Jalalabad and had gone through various points between Jalalabad and Peshawar and had been missed. We had missed them in a couple of the transit sites. We had missed them at the Khyber Pass. But finally we were able to pick them up in the refugee camp and

the kids were vaccinated. There are a lot of
these families who are going back and forth
and I think we're getting a much better grasp
on who they are.

The campaign also shifted its focus: it stopped saying that
it had achieved a ninety-eight per cent vaccination rate; instead, it
focused on the two per cent who had been missed. 'What kids are
we missing? That should be the first question that we ask in any
review meeting. Why have we missed them – and what's the plan
to reach them?' Petersen asked. It became clear that what really
made the difference was the relationship between the polio team
and the Pakistan military. Access and information crucially
depended on coordination with the army and security services to
reach those hard-to-reach children. Overall, Petersen, said, the military had become a 'much more important player in the programme
within Pakistan'. And what a difference that military intervention
made. In the course of a year the ban on vaccinators, which had
posed such an insurmountable hurdle, melted away: 'It's irrelevant.
For the most part, nobody's ever talked about it again. It's gone
away.'

The price for improved access, however, is that the polio campaign continues to be dependent on the kind of politics it longs to
be free of. Petersen admitted that the campaign is now working in
'a very complex geopolitical environment', particularly in the eastern region of the country, where some of the Pakistani Taliban
have migrated:

There is still the Taliban and there is now the
presence of ISIS in some of these areas. That's
not to say that the programme can't function

in these environments, because in Afghanistan they've actually done quite a good job in being able to negotiate and work around it.

Startlingly, in northern Syria, the polio programme is actually working with ISIS, which has allowed polio vaccination campaigns to operate in the areas it controls:

> In northern Syria, ISIS, for the most part, has supported the campaigns. They've tried to establish themselves as an administrative regime and they want to provide some credibility. The local medical councils have met with ISIS and they've agreed that teams can go in and vaccinate children. There hasn't been any barrier whatsoever.

It remains to be seen whether ISIS in Afghanistan, which often consists of former Taliban who have moved to wave a different flag, will do the same. In some cases in Afghanistan, fighting between the Taliban and ISIS has created new opportunities for vaccinators, forcing families out and into Jalalabad, making them more accessible.

At the WHO, in Geneva, Hamid Jafari said there had been breakthroughs on multiple fronts. 'The start of the military operations in specific areas of the country has dislodged some of the militant groups and that has changed the security environment and our access to a great extent.'[33]

With greater access, the number of polio cases in Pakistan fell from three hundred in 2014 to only fifty in 2015 but the situation remains tense and bloody. On 30 January 2016, at least fifteen people were killed in a suicide bombing outside a vaccination centre in Quetta; many victims were police officers guarding the clinic. The BBC reported that an officer who survived the blast said his team had been preparing to leave for neighbourhoods around Quetta when they were hit. 'Suddenly there was a loud bang and I fell to the ground, I could not see anything, there was dust everywhere,' said Shabir Ahmed, who suffered shrapnel wounds to his stomach, hands, legs and feet. 'Then I heard people screaming and sirens of ambulances.'[34]

The battle in Pakistan is far from won but as the situation has improved Jafari said that attention has moved back across the border to Afghanistan:

> Afghanistan has had a recent transition in government but I think the new government is now starting to engage and take control of the programme. That always takes a bit of time, as it did in Pakistan. But there is also the shifting security and political dynamics in the southern part of the country. A ban on vaccinations meant that half a million children had not been vaccinated since June.[35]

This ban was not ideological but was caused by human error, when health workers failed to properly administer the measles vaccine and several children died. 'A full investigation was done and the problem was identified,' Jafari said. 'The community has been told the reasons for what happened and compensation was given. It looks like it's moving towards resolution.'

Getting to the finish line in Afghanistan and Pakistan is critical but sometimes that finish line seems to inch further and further into the distance as new problems arise and have to be overcome. Afghanistan continues to exist in a state of general insecurity, with many areas not under government control. Cooperating with the Taliban in these areas makes it even more important for polio to be framed as a humanitarian issue, not political subterfuge.

To meet the 2019 date for complete eradication set by the GPEI and its donors, Petersen and his colleagues were racing towards a self-imposed deadline of stopping transmission by 2016. Tim Petersen said, 'I think there's a really good chance we will stop transmission.' With cases falling to fifty, 'that puts us in a good position to stop transmission in the low season next year. When I look at India we had forty-two cases the year before we stopped transmission. Then we went down to one and then we were done.' He smiled, a little wearily. 'So we're going to try.'[36] In large part, the team remains optimistic because of events thousands of miles away in Africa.

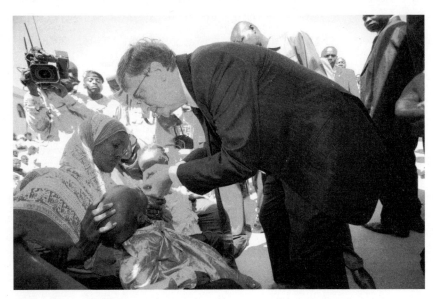

The billionaire philanthropist Bill Gates is one of the latest, but biggest, players in polio eradication. Through the Bill and Melinda Gates Foundation, he is also committed to malaria eradication and the elimination of neglected tropical diseases.

4

BILL GATES AND THE FINAL ONE PER CENT

Africa has been polio free for two years . . . we are within reach of wiping polio from the face of the earth forever.
Bill Gates[1]

In Africa, it started in the dark. By the time Bill Gates's plane touched down in Sokoto in northern Nigeria in 2009, his team had prepared intensively for his visit to the region, where polio remains endemic. After disembarking in what was once the heart of an ancient caliphate that established Islam in this part of West Africa, Gates began the rest of the journey to the home of the Sultan, the spiritual leader of northern Nigeria's seventy million Muslims.

'It was a cacophony of sound, because they brought out all the drums and the horns to greet us,' Michael Galway, deputy director of the polio programme, responsible for Africa at the Gates Foundation, remembered. 'Bill got out of the car and was ushered into the palace. He sat down on the side throne to the Sultan's throne and we waited for the Sultan's procession to come into the room.' Sultan Muhammadu Sa-ad Abubaker III entered, dressed in huge white robes, as the people called his praise. He sat down – and the lights went out. 'The Sultan looked at Bill Gates and said "Welcome to Africa".'[2]

Several years later, on 25 September 2015, the WHO announced that polio was no longer endemic in Nigeria, marking the first time that the country had interrupted the transmission of the wild virus. Laboratory data confirmed that it was more than one year since any new polio cases had been identified. The African continent was one step closer to being certified free from polio.

While health workers remained cautious about celebrating too soon, this was a remarkable achievement, given that only three years earlier Nigeria had accounted for half the world's polio cases and remained the engine of infection for West Africa and beyond. An unprecedented effort by all partners in the polio programme – and the intervention of two of the richest men in the world – mobilized 200,000 volunteers to vaccinate 45 million children under the age of five, driving down cases by ninety-two per cent.

A year earlier, in the summer of 2014, that day had seemed a long way off. In March 2014 the World Health Organization had admitted that the final stage of eradicating polio was proving to be by far the hardest, announcing: 'The fight to eliminate polio is now imperilled . . . by insecurity, targeted attacks on health workers and/ or a ban by local authorities on polio immunization.' Polio was returning to places where it had been eliminated and the risk of it spreading further remained high 'particularly in central Africa, especially from Cameroon, the Middle East and the Horn of Africa'.

By camel, truck or canoe

In his small, neon-lit office in the CDC in Atlanta, Georgia, incident manager Greg Armstrong pored over a sequence of red DNA data, laid out like a long family tree. The data represented the genetic make-up of a polio sample that had been spinning in a test tube just hours before. The sample was analysed in the laboratory of Cara Burns and her colleagues, a few hundred yards from Armstrong's office, a couple of floors up from anthrax and next door to the bland-looking rabies lab. Although the CDC, set in the leafy suburb where Hollywood had once filmed *Driving Miss Daisy*, might look like a high-tech office park, it holds a stockpile of some of the world's most lethal diseases, including smallpox, and is the nerve centre for the final stages of the global fight against polio.

Samples of African polio cases from outbreaks in Nigeria and the Horn of Africa are flown across the world to be analysed in this laboratory in an Atlanta suburb, enabling scientists to identify the exact strain and origin of the disease. 'We just got this,' Armstrong said. 'We sequence every virus that's picked up and we just recently picked up a case in Somalia, in Puntland, right about here.' Armstrong points to one of the maps of East Africa that cover the back wall of his office. The other wall consists of floor-length glass windows facing into the main Emergency Operations Center 'mission control', where the polio response is coordinated from banks of desks and computers that face a wall of screens flashing maps of polio outbreaks, information about team deployments and breaking news. Armstrong turned back to his print-off and ran his finger down the list of DNA data for the latest polio case: 'This came after about five months of not detecting cases and the question is, where has this virus been for the last five months? Sequencing is one of the few clues we have to that'.[3]

Armstrong's title, incident manager (a US fire service rank, introduced after Hurricane Katrina to denote the person in charge of disaster response), indicates the importance of his job; his gaunt and hollow-eyed appearance indicates its stresses. At the CDC, even as Middle East Respiratory Syndrome and Ebola outbreaks are urgently discussed by high-ranking officials kitted out in military dress whites, polio eradication remains a top priority. Failure would be a massive blow in terms of funding, confidence and prestige. The team worked against the clock to identify whether the virus was related to the Somali region of Ethiopia, which would have been particularly troubling as the quality of the vaccination campaign there had vastly improved recently, or whether the virus was part of a larger cluster further south in Somalia. 'The answer is it's more closely related to a number of specimens down here,' Armstrong said as he pointed to Somalia. 'So this doesn't give us any definitive answers to where it's been but the programme implications for it are that we really do need to focus the response on Somalia.'

Although Somalia managed to stop endemic polio in 2002, the country suffered from explosive outbreaks in 2005 and 2013, when the virus was transmitted from Nigeria and West Africa. Very low child vaccination rates and the breakdown of the health system caused by the twenty-year-long civil war led to huge outbreaks that the polio programme found difficult to control. 'Conducting mass vaccination is very expensive and when you have insecurity the cost doubles or triples,' explained Dr Mulugeta Debesay, who runs the polio team in Somalia:

> There is the issue of logistics. Sometimes you may have to charter flights. There is also a security challenge; you have threats against your staff and you have to decide whether to send people out and push them beyond the limit, because you are risking their lives. There are parts of Somalia that are relatively better but there are areas in South Central where you cannot even travel and there are

areas where the infrastructure is completely destroyed. Sometimes the only way to go in there is using local staff and trust them to do their job by remote control, so to speak. Then it takes a lot of time to monitor the quality of the campaign; so it takes a year to stop any new importation and almost ten rounds of mass vaccination campaigns.[3]

Brigitte Toure, senior regional immunization adviser at UNICEF for the East and Southern Africa Regional Office, explained that the outbreak in the Horn of Africa started in the Somali region, which covers three countries: Somalia, Kenya and Ethiopia. 'They don't really care about borders. People are travelling a lot, some are pastoralists but others are people who are moving for thousands of reasons, because of security, because a school is opening, to get food distribution or for social reasons to visit their parents,' she said. Key lessons learned in Somalia, Toure said, were the establishment of transit post teams and the use of a method called 'zero doses':

When we are immunizing a child we ask their mother if the child has ever received a polio vaccine. If they have never had one, we call this child a zero dose. We monitor the quality of the campaign and if it's good there is a lower number of zero doses but if you open up some inaccessible areas, you will see an increase of the zero doses.[5]

In her time working in Africa Toure has seen vaccines transported by 'car, boat, canoe, bicycle, motorcycle, donkey – even camel'.

After arriving at a local port, the vaccine is moved by truck to a local health centre, where it is kept in ordinary-looking rooms in a series of gas-powered domestic fridges. From there, it might be strapped to the back of a motorcycle and driven for miles over waterlogged and rutted rural roads. After that, it might be carried in a container slung over someone's shoulder, as they trek on foot through dense undergrowth to reach a village. In such circumstances preserving the 'cold chain' that keeps the vaccine at the correct temperature is a nightmare:

> You need to calculate the volume of your vaccine, the volume of your ice bag, the number of vaccine carriers that you will need for the area, the number of fridges or freezers and the capacity of each. It's a tedious process and on top of that it's key to have a very detailed plan of distribution of vaccines, to avoid wastage. In town it's very easy because you have fridges but when it comes to rural areas then you have to calculate the time of transport, how long it will take, how much you will need and the rate that you're at.

In Toure's opinion, countries divide into: 'Places where there are polio cases and the difficulty is getting the government on board to organize a quick response and countries where the government is totally disorganized but they officially have no cases.' She believes that reports of zero cases from those countries arise from the fact that the campaign can't get in to undertake surveillance and identify cases, rather than a true absence of polio.

Victory in Nigeria

In Atlanta, as the CDC team sweated through the hot southern summer of 2014, it kept a sharp focus on Africa. Cases had been significantly reduced in the Horn of Africa and the team was optimistic about Nigeria, where progress was clear. Overseeing the polio campaign in Africa for CDC, Craig Allen said:

> Northern Nigeria is clearly our first priority in Africa. If we can stop it in Nigeria, Africa will be polio-free very quickly. Vaccination rates have now risen to over eighty per cent in northern Nigeria, the highest ever. Last year at this time we had twenty-two cases of polio, this year we've got three.[6]

The final push is illustrated by the population density maps and quality assurance surveys from northern Nigeria that paper the walls of Greg Armstrong's office: 'Green is good, red is bad,' Armstrong said, pointing to the charts showing the level of vaccination coverage rising and falling. Rising figures for vaccination coverage show an interruption in 2012, when vaccination workers in Borno and Kano were targeted and killed by rebel groups in a series of attacks.

For more than a decade, Nigeria struggled to control polio after some northern states imposed a year-long boycott of the vaccine in 2003. The vast and populous country, with 155 million people spread across 36 states, first instituted a polio campaign in 1996, at a time when Nigeria was in a state of deep economic and political crisis. Under the dictatorship of General Abache, immunization was transferred to under-resourced local authorities and vaccination rates dropped as low as thirty per cent. In addition, General

Abache's wife was accused of corruption involving vaccine procurement and the director of the immunization programme was eventually accused of gross incompetence and corruption. The transition to civilian government after the death of Abache in 1998 raised new hopes for immunization programmes, and polio vaccination in the government-dominated southern part of the country gained ground.

Soon, however, the vaccination campaign was a source of enmity between the southern, Christian, part of the country and the Muslim north. Some state governors and religious leaders in the predominantly Islamic north alleged that Western powers contaminated the vaccines, to spread sterility and HIV among Muslims. Why else, they asked, were polio vaccines free and readily available when other – more needed – medicine was both expensive and scarce? The leadership of the Muslim umbrella group Jama'atul Nasril Islam and Nigeria's Supreme Council for Sharia called for a boycott. The states of Kano, Bauchi, Kaduna and Zamfara suspended polio vaccination pending an investigation into the vaccine's safety. Families (who had good reason to mistrust vaccine programmes; in 1996, a disastrous meningitis trial led by Pfizer left eleven children dead) began evading the vaccinators, marking their doors with signs that falsely indicated they been vaccinated or painting their children's fingers to imitate the mark of the polio team.

The polio campaign countered with a huge effort to mobilize a response to the 'refusers'. The partner organizations and the US government lobbied senior Nigerian political and religious leaders, the Organization of the Islamic Conference and the African Union to resolve the situation. The polio programme bought vaccine from an Islamic country, Indonesia, and asked Saudi Arabia to require vaccine certificates from all Muslims attending the Haj.

Their efforts were successful and in 2004 the ban was lifted. The then state governor of Kano allowed President Obasanjo to personally administer the polio drops to the governor's one-year-old daughter.

Support for polio vaccination was furthered strengthened when Muhammadu Sa-ad Abubaker III, who became Sultan of Sokoto in 2006, voiced his support for the campaign. The sultan acceded to the throne after his brother died in a plane crash; before that he was a high-ranking soldier and had spent much of his thirty-one years in the army as a peacekeeper, as well as a Nigerian defence attaché to Pakistan and Afghanistan. In other words, he was a man who understood how the world worked. Described by a fellow officer as 'strong-willed, blunt, a very good soldier, a strict disciplinarian and a man who has strong opinions on all issues', he was a man the polio campaign could do business with. According to Michael Galway, the trip he made with Bill Gates in 2009 was a key moment in the process:

> His direct engagement did a couple of things. One, while he was there the state governors signed an agreement to meet certain commitments to the programme and that document became the blue chip commitment which is still in place today. The second big part of his trip to Sokoto was his engagement with the Sultan, as well as the first class Amirs, and they told him that they were ready to take on a more direct operational role in the programme. You may recall that we weren't that far removed from the period in which Nigeria had actually stopped immunizing

kids because of fears of the safety of the vaccine. So the Sultan formed a committee called
the Northern Traditional Leaders Committee
on polio eradication and that committee was
another strong institution within the polio
programme.[7]

The sultan's authority extended his influence down to the smallest
hamlets, making 2009, in Galway's opinion, a turning point in the
programme, in terms of political engagement and getting communities to participate.

The success of working with the community and religious
leaders in Nigeria led the GPEI to begin working with Islamic
scholars to establish an Islamic Advisory Group (IAG) to engage
with the Islamic community across the world, particularly in
Pakistan. 'The Qur'an says that protection of children is one of
the most important responsibilities of parents for any child,' said
Dr Yagoub Al Masrou, a Saudi Arabian epidemiologist, one of the
key figures in setting up the IAG.

In March 2013, Islamic scholars met in Cairo to found the IAG.
Representatives came from different agencies, including the
Al-Azhar Al-Sharif from Egypt, the International Islamic Academy
from Saudi Arabia, the Islamic Development Bank and other
organizations from Islamic countries, with technical support from
both WHO and UNICEF. Before the end of the year the group
drew up the Jeddah Declaration, stating that polio vaccines were
safe, and compliant with Islam. Al Masrou declared that:

In many instances misconceptions about the
Islamic religion were used as obstacles to
polio vaccination. Islamic scientists analysed

the composition of these vaccines and proved
these claims to be false. We wanted to pass
that on to the religious community and the
local community at every level.[8]

The Jeddah Declaration affirmed the importance of solidarity
among Islamic scientists and religious leaders in combatting polio
and support for global polio eradication efforts. Also the group
recommended that Islamic countries that had already got rid of
the disease should support those countries where polio was still
endemic.

Tackling 'the refusers' was a big step forward in wiping out
polio; the polio team had taken on entrenched ideas and myths and
won. On the ground, however, there were more practical problems.

The partners in the polio programme had worked hard to
improve polio vaccination in Nigeria, introducing a series of
regional emergency operations centres (like the one in Afghanistan),
satellite mapping of more than one hundred thousand communities, drastically improved data collection and surveillance, using
mobile phone apps to track vaccinators and make sure they were
turning up at their destinations, and wrapping polio vaccination
efforts within bigger health campaigns that included routine vaccinations and maternal health checks in a 'one-stop shop' for locals
who were tired of the endless polio rounds. More than two thousand female volunteers were trained for crucial roles, becoming
the lynchpins of the campaign as they visited all the houses in
their village and talked to mothers about polio, immunization and
other problems.

One female 'community mobilizer', thirty-year-old Binta
Barau, described how she had been to more than one hundred
houses in Marke Village, Bunkure to check that all children under

five were vaccinated. 'Convincing families and mothers has not been an easy task,' she told UNICEF. 'Parents queried why we are delivering polio immunization services every two months when they still had to grapple with other problems like malaria and diarrhoea.' One father, 44-year-old Sabiu Abiola, said: 'By the grace of God, my children are all healthy, so why do you want to administer medicines to them?' But Binta had been trained to use an illustrated flipbook that explained why immunization was important. It worked; the man relented and allowed his child to be vaccinated.[9]

From 2010, a huge effort built up some of the programme's structures, Michael Galway said:

> The work of mapping all of the communities in the north was a big intervention, because we identified areas we weren't reaching to begin with. The people who lived there knew they existed for sure but the polio programme was somewhat agnostic about them. There were thousands of hamlets. Everybody can find a village of more than a hundred households but when you get down to small hamlets, especially in nomadic areas, ten, twenty houses stretched out over a kilometre, then you start to fall off the radar. The polio programme was able to put those back on the radar through GPS mapping and then make sure that those places were included in the programme's plan and got vaccines. The programme is really an exercise in anthropology and ethnography when it gets down to it.[10]

As advanced as the science was, the success of the polio programme rested on old-fashioned explorers: men and women charting unmapped areas, travelling off road for days, reaching villages no outsider had visited, and discovering who those people were, what they believed in and what would make them agree to immunize their children.

The campaign also set up an Emergency Operations Center in northern Nigeria, modelled on the EOC at the CDC in Atlanta. 'Until about 2010, the programme in Nigeria was working from various locations,' Michael Galway said. 'WHO had their offices, the government had its quarters and this was replicated at the state level. We just weren't getting the efficiency and the effectiveness of working together.' This simple concept would be rolled out elsewhere, such as in Afghanistan: bring people together in one building, agree how they would work together and how much time would they spend apart, divide functions into different categories of work (an operations group, a strategy group, communications group, a data group) and set up a rhythm for reviewing the programme together in real time. 'It was a very successful intervention in Nigeria. It really did help us pick up our game and push to another level of coordination and effectiveness,' Galway said. The EOC was the 'place where people get bad names, you know; this area's not performing well, who are they, why are they not performing well, let's bring them together, let's sit with them and find out what's going right, what's going wrong'.

Drawing on shared resources and time spent together, the programme started to innovate. One innovation was the introduction of activated polio vaccine contained in easy-top-carry canteens, which sustained the immunization of four and a half million children over two years. At the CDC, Greg Armstrong pointed to the population density map for Kano:

The improvement has been particularly big
in Kano, which is extremely important. Kano
has been the engine of polio transmission for
all of West Africa for years, probably because
it's so populated. The denser the population
and the poorer the sanitation, the higher the
coverage we have to get to actually interrupt
transmission.[11]

Michael Galway counted off a variety of strategies the polio
programme introduced to bring transmission down to zero. 'We
introduced a whole new operational approach in terms of immunizing children on the street outside their homes, because we felt
that we were missing children inside the houses. So how could we
get children outside and get them immunized?' There was a suspi-
cion in the programme that one of the worst case scenarios was
actually happening: some vaccinators were going into households
and not immunizing the children but nonetheless marking their
fingers as if they had:

So the programme came up with a strategy in
which they would immunize kids on the streets
for two days before they would actually then
go and knock on the door and go inside the
house. It was a very successful strategy because
there were some very positive enticements to
bring kids out of the house. Sometimes they
used clowns, sometimes they used incentives
like soap for mothers. We found that in some
areas we had a high number of kids who had
never been immunized because they were

recorded as resistant or we found out that
there were a lot more kids in those households than we suspected. Now seventy to
eighty per cent of kids were being immunized
on the street through that campaign.[12]

Kano instituted another strategy called the 'End Game', in which vaccinators listed every single child whose parents had refused immunization and passed the list to the district head, who then went with local leaders to talk to the family about their reasons. But not everyone was happy. On a trip to the region in 2014 the health writer Alex Kornblum noted that:

Ordinary Nigerians told me that what
angered them most about the polio campaign
was that the government wasn't doing anything else for them. Nigeria has the largest
economy in Africa and yet its people are
among the poorest in the world. Although
roughly $52 billion from oil sales is earmarked
annually for poverty reduction, more than
seventy per cent of the population lives on
less than a dollar a day.[13]

It was a complaint with some justification: polio cases were falling but Nigeria's health record for immunizing children for other diseases was not only one of the lowest in the region but fell throughout the 1990s. In 2003, only Sierra Leone, mired in years of civil war, had a lower rate. Delayed government payment meant vaccines were delivered too late or not at all and half of the children in northern Nigeria were malnourished.

In other words, the polio campaign had to do a lot more than target polio if it were to succeed in Nigeria. In a hugely expensive step, the programme started 'health camps' to tackle some of the country's wider problems and deliver other services. This was important because, just as in India, a sense of frustration with the polio programme was growing on the ground. Why was the polio vaccine the only government service being delivered, people asked, when so many other things were needed?

'The Health Camp strategy . . . was a difficult proposition because of the cost and because of the scale of intervention that was required but it was successful,' Michael Galway said. Kano State set up two thousand health camps and treated one and a half million people in six months. 'That engendered a lot of goodwill for the programme; people felt like we were here to address things that are important to them, including basic health services, diagnostic services and also referrals.'

The improvements paid off: by December 2013 Borno had reduced the number of children who couldn't be reached by vaccinators from sixty to thirteen per cent. Yet in northern Nigeria security remained the biggest problem. Only a small number of provinces were unsafe but they were the polio hotspots. 'The big threat in Nigeria is still terrorism, like Boko Haram,' Craig Allen said. 'There are places that are just not safe and so we're struggling there. We're trying to do campaigns at the edges of the insecure areas. We have posts at the border; when people come and go we vaccinate them.'[14] Liaising with the Nigerian army for the latest information on rebel movements, teams of vaccinators swoop into an area at short notice, vaccinate quickly, within hours, and leave before the area becomes unsafe.

'We have thirty-six states and three of them are under emergency. In the other thirty-three states there is absolutely no problem,' said the eminent Nigerian virologist, Oyewale Tomori:

> The uncertainty of where insurgents will strike next has actually grounded activities in some places, such as many airlines don't go to Maiduguri because they are not sure what is going to happen. So because of the uncertainty regular things are not done and you have to have occasional checks, working with the security people to see which parts of the state is safe and then go in there and do what you can do before they are alerted and then you have to get out.[15]

Twelve teams of vaccinators, each with three to four members, sweep through an area, working as quickly as possible, in a 'hit and run' operation. As well as setting up health camps and vaccinating in camps for internally displaced people across the north east of Nigeria, the campaign started what it called a 'Firewall Strategy'. 'That means immunizing in border areas outside no-go zones, so that if people come out of those areas they get picked up,' Galway explained. 'There's a lot more immunization around local government areas that are out of bounds or under the control of Boko Haram.'[16]

Adapting a strategy used in Afghanistan, the polio programme set up a team of what they call 'Permanent Vaccinators'. Instead of sending a team to a particular location for a number of days, they take the polio programme completely off schedule, to allow vaccinators to work quietly and unobtrusively, delivering vaccines where and when they can. 'We say "this is your catchment area and in the course of the month, whenever you get a chance, immunize". It could be at night, could be early morning . . . it's off grid.' Of course, local people cannot operate a 'hit and run' or a 'firewall' strategy and it is they who must live with the consequences. In 2014, female polio vaccinators reported that they had been 'stoned, cursed, doused with hot oil and had guns pulled on them. In some areas, they hide their UNICEF thermos bags under their hijabs.'[17] 'People demand vaccination in most of these areas,' Oyewale Tomori said. But he was far from confident Nigeria can remain polio-free unless the Boko Haram insurgency is comprehensively beaten: 'We finally got our act together after so many years and now there is hope for Africa. We have to make sure polio does not come back.'[18]

Things were better, Michael Galway admitted, but:

> We're paying a significant price. The total number of children in Borno State, eligible children for the polio programme is about 1.6 million kids under the age of five. About this time last year about a million of those kids were not being immunized consistently because of insurgency, violence and fear. Vaccinators were just too afraid to go and who can blame them?

Galway's best estimate is that 250,000 children in Borno State were not being reached by the polio programme and hadn't been reached for more than a year. 'That's a huge number of under-immunized kids. That's a huge vulnerability for polio eradication in Africa.' Borno State has a low level of immunization against other diseases, so 'if a virus makes its way into Borno that's a huge problem'.[19]

If the success of the polio programme in Nigeria highlighted a nimble dexterity in bringing together the lessons learned in India, Pakistan and Afghanistan, it also marked a ruthless turnaround in management and efficiency prompted and funded in large part by the intervention of the Bill and Melinda Gates Foundation, and by the personal interest of Bill Gates.

'Deep in the weeds' with Bill Gates

He might be a billionaire based thousands of miles away but his team, as well as the Nigerian health workers, have become used to the fact that he wants to meet them, talk to them, cajole them, get them on conference calls to ask why things weren't working or have gone wrong. 'They all have a lot of respect for Bill,' Galway said:

> I think they're all so taken by the fact that this person who's so far away is interested in their kids and I think that they also think that it's just not acceptable for someone to come from the United States and be more interested in taking care of their children than they are. They believe that they should be taking care of them themselves. And they did.[20]

When Bill Gates met Africa's richest man, the Nigerian commodities tycoon Aliko Dangote, they discussed how other philanthropists could get involved in the programme. 'Bill and Mr Dangote hit it off right away; I think they're really kindred spirits,' Michael Galway said. 'Bill followed up and we started to work with him in a much more direct way.' In a good-cop bad-cop routine, Gates usually plays the understanding partner, while Dangote comes in with a tougher line, demanding that things get done.

Dangote comes from a wealthy Hausa Muslim family in Kano. He used a loan from his uncle to establish a commodities and trading business in 1977 and built a profitable business in cement. By 2015, the Dangote Group was the largest business conglomerate in West Africa, dominating telecoms, textiles and commodities such as cement, sugar, salt and rice, as well as being the largest private employer in Nigeria. Employees tremble in Dangote's presence. A telephone interview for this book went through a string of nervous intermediaries before the great man finally came on the line to outline his healthcare vision in person.

'He cuts much more closely to the bone when he talks to governors and traditional leaders and the president around what it is that the country needs to deliver for their own children,' Galway said. 'He's definitely of the school that this is a job for Nigerians to take care of themselves and was very forceful with that message. On polio he has played a special role in getting the country to where it is today.'

One of the big issues the two men had to tackle was funding and corruption. British development officials reported in 2011 that trying to find missing money was a 'fool's errand'. They discovered that 'the majority of clinics in one state kept no budgetary records at all and most of those that did spent far less than state officials claimed'. Poverty alleviation funding allocated for basic resources disappeared behind a bewildering smokescreen of shell companies and scams. One reporter investigating a health clinic found it doubled as a dress shop, with key medicines out of stock, and the nurse on her way to another clinic because she thought she had malaria.

If the polio programme were to be successful, it would have to find new ways to pay for itself no international institution could have attempted. In a huge step, the Gates Foundation negotiated loan restructuring agreements with the World Bank, along the lines of a buy-back scheme. In essence, the Foundation agreed to pay off development loans of US$65–$95 million given by the World Bank to countries such as Nigeria if those countries met their polio eradication targets. On top of that, the polio programme had to get to grips with the funding problems that beset routine immunization programmes, Galway said:

> We've used our funding to try to drive perfor-
> mance and for innovation as well as the nuts
> and bolts of the polio programme. In Kano
> State in late 2012 and then Bauchi State last
> year, we entered into a three-way partnership
> between ourselves, Dangote and the two
> state governments to support a routine
> immunization programme. We were very
> clear from the outset that there was no such
> thing as a Gates project on immunization or a
> Dangote project on immunization. We were
> there to support the state governments to
> fully fund and run a well-executed routine
> immunization programme that would be
> very unlike a campaign. Polio in many ways
> is a campaign model, routine immunization
> is forever.

First, the programme identified the true costs of routine immunization; no easy task for something so chronically under-funded. This became a three-year plan, in which the two foundations agreed to absorb seventy-five per cent of the cost in the first year, fifty per cent of the cost in the second year, twenty-five per cent in the third year and hand over to the government in the fourth year. 'It's been a great adventure,' Galway said, diplomatically:

> When we started in Kano we had an immun-
> ization rate of thirty-five per cent; this year it's
> up to eighty per cent. So we've transformed
> the vaccine supply. You can get routine immun-
> ization now in just about every area you need
> in Kano State. We've got funds in the health
> clinics so that vaccinators can go straight out
> to the field and immunize in those small
> hamlets. There are lots of problems along the
> way, I'm not saying it's the panacea but it's
> shown a great collaboration between our two
> foundations and the government.'[21]

'What we're actually doing is putting money directly into the state budgets,' said Vio Mitchell, of the Gates Foundation:

> For example, all thirteen thousand of the
> health staff in Kano State were mostly
> employed under various different local minis-
> tries and the ministry of local governments.
> But you had the State Primary Health Care
> Board, which was actually responsible for
> delivering primary healthcare. So one of the
> things we said as a prerequisite for our
> engagement was that all the health staff had
> to be transferred under the State Primary
> Health Care Board. People said it will never
> happen but it did.

Corruption or 'leaking funds', as Mitchell called it, was a big problem. 'You would release, say, the equivalent of $100 and by the time it got down to a health facility, it might be down to two dollars but they would be asked to report spending at least $75.' A system of leaking funds and false reporting created a system 'that just never went anywhere'. The Gates Foundation agreed to pay money directly into local bank accounts. 'Now we're making sure that the money is getting to the facility to pay for fuel, to pay for outreach, the things that are so important to make a fundamental immunization primary healthcare system work. That just wasn't happening before.'[22] While some state health workers seized the new opportunities, Mitchell says others were more wary and needed much more training and persuasion.

Mitchell paints an optimistic picture but those familiar with aid and development in Nigeria were far from convinced of the success of the strategy. 'The program was no match for Nigeria's legendary corruption', Alex Kornblum wrote:

> When I was there, I heard a new story every day: shortly before I arrived, some of the vaccinators were caught throwing out their vaccines and then reporting preposterously high success rates to their bosses; then local government officials were caught reselling toys and biscuits intended as gifts for vaccinated children; then I heard that officials from one of the international agencies were going around with sacks of cash, for who knows what purpose.

The polio campaign had reached a critical juncture. The more intense intervention of Bill Gates gave it fresh impetus but it also increased pressure and brought moments of extreme tension. In one memorable incident, a group of angry young men took over a vaccine store and refused to release the supplies until a local politician delivered on a three-year-old promise to provide electricity. 'When the police turned up, hundreds of women and children formed a barricade, chanting, "If you want the vaccine, you'll have to kill us first!"'[23]

Apoorva Mallya had been involved in the polio programme at the Gates Foundation since the first request for funding hit his desk on his third day in office. The request was from Rotary, asking for a big matching grant. Bill Gates was interested but never very motivated to just give money. According to the *Rolling Stone* writer Jeff Goodell, Gates's philanthropy was driven by the belief that 'the world is a giant operating system that just needs to be debugged; he thinks there's an app for everything, from ending global poverty, to improving education, to fighting disease and climate change. He has no patience with anyone who says any problem is too complicated.'[24] What more could he do, Gates wondered, to make polio eradication really work?

'That essentially set off a journey with a few people, including myself, to look across the global community, see what's going on with polio eradication and see how the Gates Foundation might help.' Mallya said. The team grew and the programme got bigger:

> We were learning how to become sophisti-
> cated in being a polio eradication partner and
> we were not sophisticated at the beginning.
> We didn't want to come in and break every-
> thing, especially when ninety-five per cent of
> the things were working. We wanted to come
> in and figure out that last five per cent.

Figuring out that last five per cent left the polio strategy 'deep in the weeds,' Mallya said, adding that the Gates Foundation pushed it towards innovation:

> What the global community had seen, before the Gates Foundation got involved, was a strategy adopted in the late eighties/early nineties, perfected in Latin America and then expanded to the global effort. It was a strategy that worked fairly easily in almost every single country in the world and made incredible progress but the last places are truly different and will not respond in the same way. Much of the work was down at the micro level: to see in a specific village why children were not being reached with the vaccine. That's something that the global programme never had to wrestle with before and that's what I think we helped contribute to building up, that expertise around detailed planning and detailed implementation.[25]

Before the Gates Foundation became closely involved in the polio programme even those closest to the campaign noted a sense of fatigue and stalemate. Cooperation between agencies sometimes left much to be desired. Data collection could be patchy. Government support ebbed and flowed. In the opinion of Sir Liam Donaldson, former chief medical officer for England and chief medical adviser for the UK, polio eradication had stagnated for ten years, unable to tackle the crucial final one per cent of cases. 'The reasons weren't technical problems, they were people problems,' Donaldson said.

Taking over as chair of the newly formed GPEI Independent Monitoring Board, with oversight of the campaign and all its partners, he quickly identified problems with leadership and issues with frontline vaccinators not being able to reach communities and discuss with families why the vaccine was needed. One thing that astonished him was that until the monitoring board was formed there was no figure for how many children were missed from the vaccination campaign. Or rather, that a figure did exist but was deeply buried in a report under a different heading. The figure Donaldson identified, and prominently displayed, was 2.5 million. 'Two-and-a-half million missed children. How are you going to eradicate the final one per cent of polio when you have 2.5 million missed children?'[26]

Jay Wenger, who now leads the polio programme at the Gates Foundation after working at WHO, says the main contribution of the foundation was to be 'a major proponent of innovation and ensuring that we use the latest thinking and the most aggressive tactics to actually get the job done'.[27] Such focus and aggressive tactics, however, ruffled feathers in the larger partner organizations that had worked on polio eradication for decades.

'At the time the polio programme was really struggling for money. So the engagement of the Gates Foundation was a lifeline as far as financial support was concerned,' Chris Maher, from the WHO, said. 'It also brought a different sort of energy to the programme and a different level of expectation. On the positive side fresh eyes were looking at it and on the less positive side, those eyes weren't necessarily used to seeing the sorts of issues that we were facing.' Maher concluded that:

> Overall it would be hard to say that it wasn't a
> good thing, it was a very good thing. But of

course when you have any big new partner joining an initiative saying 'hey this is great, we really want to get into this and we want to see things done faster', then it creates a certain set of ripples that you've got to cope with.[28]

Carol Pandak, who heads Rotary's polio programme, PolioPlus, agreed:

> They're a big new player that came to the table a little bit later in the game and there had to be some adjustments made to allow for the extraordinary funding that they have available and the intensity with which they approach anything that they do . . . The Gates Foundation has the voice of Bill Gates and having the voice of Bill Gates has been helpful.[29]

No one disputes that the voice of Bill Gates increases scrutiny and adds pressure. Vio Mitchell described the six-monthly conference calls with Bill Gates, Aliko Dangote and the team in northern Nigeria. 'Bill is totally into them,' Mitchell said, adding:

> We had one really hot moment. This was about a year ago with the Governor of Kano. Bill and Dangote love to hear about problems. That's how they thrive. They don't want to hear everything's rosy. But one of the things the commissioner had talked about during

one of the reviews was that we were still slow on getting some of these bank accounts for the health facilities functional because there were problems getting signatories and making sure the right people were signing. And there were also health facilities that were having problems with what they call retiring a fund or accounting for their funds.

Mitchell said there were no 'nefarious' reasons; it was just that people had never done it before and needed to learn how:

> Bill was like, 'oh that's fine, we'll work on it for next time'. And Dangote said 'well, we need to get our act together here'. But the governor was just apoplectic and after the meeting he just reamed everybody out. I was there in Kano at the time and he said to me 'You can ring me at one in the morning, why didn't you tell me that this is a problem?'

Mitchell replied that she didn't think it was that big a problem and she didn't want to bother him but admitted that 'we do have some cultural needs that we need to manage and not everything can be a hundred per cent. It's been a journey.'[30]

That specific focus and attention to detail, was crucial, agreed Aporva Mallya, as the Gates Foundation began to work more closely with all the partners in the programme, initiated the setting up of the Independent Monitoring Board and started working on the ground in communities in northern Nigeria, Pakistan and Afghanistan:

We came in with a lot of funding for supporting existing activities but then we realized
that one of the strengths of the Gates
Foundation is we invest a lot in innovation
and a lot in high risk areas. We started working with the other partners to do some clinical
trials, some field studies of new ideas, new
innovations, new vaccines, new technologies
that may be helpful. We also realized how to
best use Mr Gates because you don't want
him going to every country every month and
saying, what are you doing about polio? He
needs to go in very targeted ways, to meet
with presidents and prime ministers and programme leaders.[31]

While Gates personally lobbied heads of government and other
philanthropists to support polio eradication (and fired off late-night
emails to his employees, suggesting new ideas and wanting to
know what was going on), the Foundation provided much of the
support for the technology behind the campaign, such as the smart-
phone app that allows vaccinators to be tracked in the field. It also
focused on the one major area requiring transformation: the
vaccine itself.

The vaccine

For years the workhorse of the polio eradication programme was
the oral polio vaccine, OPV, which was cheap, easy to administer
and provided a gut immunity to polio that spread throughout an
entire community. The problems associated with OPV, however,
were well known: it led to a small number of polio cases every year

and many doses had to be administered to the same population to ensure coverage. The former director of the US Immunization Programme, Walter Orenstein, said:

> We had about eight people a year in the US that were paralysed by the oral vaccine and I will never forget being in a room with them in '95 or '96 and seeing these horribly crippled people. That spurred our move into IPV in the US to try and avoid that.[32]

In addition, the OPV was a trivalent vaccine; it targeted all three strains of the polio virus but no wild type 2 polio cases have been reported since 1999 and no wild type 3 polio virus since 11 November 2012. 'The original vaccines that were licensed were monovalent vaccines but that required giving different vaccines at different times and different doses and so in 1963 the trivalent vaccine was licensed', Walter Orenstein explained. But health workers ping-ponged between different vaccines, returning to monovalent vaccines after polio type 2 was wiped out and the trivalent vaccine was causing vaccine-induced cases of the strain.

When Liam Donaldson joined the monitoring board a debate raged within the polio community about whether it was necessary to use OPV in conjunction with IPV, Salk's vaccine, made with deactivated virus and administered through a single, but more expensive, injection. Although it was agreed that eradication could only be achieved by using both vaccines, Donaldson found progress to a decision frustratingly slow. 'Bill Gates's first question to me was what about IPV and OPV?' Donaldson said.[33] The two men met shortly after Donaldson took the helm at the IMB; Donaldson was taken aback by Gates's mastery of technical detail and searching questioning. Gates's intervention forced the debate to a conclusion and the Gates Foundation led the development of a bivalent OPV vaccine that targeted the two remaining strains. 'They have played a big role in reconsideration of the inactivated polio vaccine and they've been very supportive of the bivalent vaccine,' Walter Orenstein agreed.[34]

John Modlin joined the Gates Foundation as deputy director for global development for the polio programme, with responsibility for research, following a distinguished career as chair of the Department of Paediatrics and senior advising dean at Geisel School of Medicine at Dartmouth. His research at Johns Hopkins University was instrumental in the decision to change polio vaccination policy in the United States in 1997 and again in 2000. Sitting in the Gates Foundation office, overlooking the Seattle Space Needle, he explained how Bill Gates feeds into the research programme:

> Bill is not a technical expert in polio but he certainly brings a different view. I won't characterize it as an outside view, because he obviously has a passionate interest in eradication but he has a very active creative mind and reads widely and he comes up with ideas that we may not come up with. When that gets fed into our agenda, it winds up affecting it and we need to be responsive to many of his ideas.

Modlin says that while Gates's ideas all have a good basis, sometimes the team has to explain the technical reasons why they might not work. 'When that's the case it's our responsibility to say so.' And sometimes, one suspects, that might not be easy.

At the moment the research team is working on developing a new live attenuated oral polio vaccine for all three strains, focusing on type 2. 'The preclinical development for that project has been going on for about four years and has now moved into the clinical phase of testing these new strains,' Modlin said. 'This was really Bill Gates's idea; to have a safer live strain that would combine the safety of IPV and the advantages of oral polio vaccine and could perhaps be less expensive to make and easier to deliver.' It is an approach that not everyone in the polio community believes necessary but Modlin believes it symbolizes the innovation that the Gates Foundation brings:

> Not everybody agrees. This is a vaccine that may not ever be used. When you tell someone who is familiar with polio eradication that you want to develop a new live attenuated vaccine, the mindset is 'but isn't that what we're trying to get rid of?' But if we did have to use it, there's no question that this would be a better alternative than the monovalent strains which we have today.[35]

More immediately, the GPEI strategic polio endgame strategic plan for 2013 to 2018 stated that the programme would switch from using trivalent OPV to bivalent OPV in April 2016 and introduce one dose of IPV into routine immunization schedules for all countries.

'One of the big elements of the 2013 to 2018 strategic plan is the introduction and use of IPV,' Jay Wenger said. The problem was that there wasn't a supply of IPV available, and it was expensive and mainly used in developed countries. To bring IPV to the rest of the world, Wenger said, the Gates Foundation worked to solve problems of financing and supply: 'The Gates Foundation has been key in finally coming to an arrangement with the manufacturers that will allow purchase of the vaccine of the IPV by UNICEF for a much reduced price'.[36] The current UNICEF tender for the biggest producer of IPV is now about at a dollar a dose, reduced from several dollars. In addition, the GAVI alliance, which provides vaccines to the developing world, agreed to support providing the IPV to low-income countries. Walter Orenstein noted:

> The GAVI price is about five times more per dose than the oral vaccine. That's still a substantial reduction and the manufacturers have given GAVI-eligible countries a major reduction in price. The prices for the middle income countries are still between two and three dollars a dose, so that will require some financial assistance . . . the other issue about IPV is nobody knows at this stage how long it would be used.[37]

The longer it takes to eradicate polio, the greater the cost. And no one really knows how long it will take.

In autumn 2015, India began incorporating the IPV into its immunization schedule, following WHO recommendations that the dose should be given at fourteen weeks of age after at least two or three doses of oral vaccine. The CDC incident manager John Vertefeuille said: 'With the certification that there is no more type 2 virus circulating in the wild it becomes really urgent that we remove OPV type 2 from the programme. Coordinating the addition of IPV into so many countries all at once over a very short period of time will be complicated.'[38] He added that the switch, which began in April 2016, is unparalleled in terms of rolling out a vaccine over hundreds of countries in a fairly small period of time.

Introducing IPV, with its logistical headaches of multiple injections and temperature-controlled storage, is challenging but it is far from the only factor that could throw the polio eradication campaign off course.

The clock is ticking

Since the spontaneous outbreak of joy that greeted the announcement of Jonas Salk's vaccine in 1953 the polio campaign has been hampered by crises big and small. It has been delayed by wars, insurgency, election campaigns, religious objections and family fears. In the course of the eradication effort an estimated ten million cases of paralysis have been prevented and more than one and half million lives saved. But the clock is ticking. Samples of the disease continue to wing their way to Cara Burns's polio laboratory in Atlanta from around the world, while the Emergency Operations Center flashes in daily updates of outbreaks and security threats that emphasize the race against time until funders' interest and commitment wane.

Throughout 2015 the team continued to deal with vaccine-derived outbreaks across Africa, Europe, Ukraine and Laos. Then, to the deep dismay of a campaign now governed by a tightest of self-imposed deadlines, and on the very day the polio campaign had planned to celebrate the anniversary of Africa being polio-free, Michael Galway's warnings about the unvaccinated children of Borno State came to horrible fruition. On 11 August 2016 the WHO confirmed that two Nigerian children in northern Borno had been paralysed by the disease. Genetic sequencing suggested that the cases were caused by a wild strain last detected in Borno in 2011, indicating the virus had circulated undetected for five years and more cases might well be reported from areas previously off limits to the government and NGOs. On the ground the news was met with little surprise among those who understood the ramifications of so many unvaccinated children. Health workers quickly sprang into action, vaccinating and monitoring and doing what they know how to do so very well. But the symbolic blow went deep.

Veteran polio campaigners, such as Hamid Jafari (who stepped down from running the GPEI at WHO in January 2016, and returned to the CDC in Atlanta), admit the task sometimes becomes overwhelming. 'It's very tough, it's very intense, it's very complicated and it's very exhausting sometimes,' Jafari said wearily. In India, then in Africa, now in Pakistan and Afghanistan, it takes nothing short of momentous belief to drive them on. 'You have this goal which is the ultimate in disease control: eradication,' Jafari said:

> And that is also the ultimate in equality. The imperative of this programme is that in order to succeed you have to reach everybody. You have to reach people who, no matter what their risk and their religion might be or even their opposing views to yours; you still need to vaccinate their children.

Polio eradication is a twentieth-century dream, conceived by idealists and driven by big international institutions and mass mobilizations of volunteers, working together to make a better world for all. It must succeed or fail, however, in a twenty-first century marked by factionalism, religious intolerance and rising inequality. Hamid Jafari shrugged as he encapsulated what he thinks makes his life's work unique: 'In an unequal world, there are places where only polio eradication has touched.'[39]

The tiny *Aedes aegypti* mosquito is the size of a grape seed but can spread yellow fever, dengue fever and the Zika virus.

5

THE PROPHET

Fred Soper was a physically imposing man. He wore a suit, it was said, like a uniform. His hair was swept straight back from his forehead. His eyes were narrow. He had large wire-rimmed glasses, and a fastidiously maintained David Niven mustache. . .
Malcolm Gladwell, *New Yorker*[1]

With so few remaining cases and so many other pressing health concerns, over forty years polio eradication morphed from a medical blessing into an ideological standpoint; one its supporters feel obliged to defend against those who argue that it is absorbing too much attention and money from donors and the global health community.

In 2011, Bill Gates was enraged when high-profile critics of polio spending emerged to say his intense personal commitment had skewed attention from where it was most needed. Richard Horton, editor of *The Lancet* tweeted: 'Bill Gates's obsession with polio is distorting priorities in other critical BMGF [Gates Foundation] areas. Global health does not depend on polio eradication.' Arthur L. Caplan, director of the University of Pennsylvania's bioethics centre, himself a polio survivor, entered the fray to say the best that could be hoped for was disease control. The smallpox veteran, D. A. Henderson, added his opinion to the debate: 'Fighting polio has always had an emotional factor – the children in braces, the March of Dimes posters . . . But it doesn't kill as many as measles. It's not in the top twenty.'

Gates responded angrily to the suggestion that the $1 billion of USAID funding for the polio vaccine should be diverted into other areas, including pneumonia, measles, meningitis and malaria: 'These cynics should do a real paper that says how many kids they're really talking about,' he said in an interview. 'If you don't keep up the pressure on polio, you're accepting 100,000 to 200,000 crippled or dead children a year.'[2] The Gates Foundation summoned leading economists to produce a robust study on the economic and health benefits of polio eradication, but it seemed stung by the opposition.

Although the campaign to eradicate polio has been greater in scale and funding than any other disease eradication programme, its ambition is not unique, nor are the criticisms levelled against it. Words such as 'campaign', 'eradicate' and 'surveillance' place it in the lexicon of a very specific type of disease epidemiology that developed in the twentieth century and exemplified a technical, autocratic and medical approach to 'health' championed sometimes fanatically by its adherents, who mainly come from the United States.

To understand how polio fits into the history of disease eradication and global health it is necessary to go back to the birth of microbiology and one of the world's first disease eradication campaigns: the fight against yellow fever.

The birth of microbiology
Bacteria had been seen through microscopes since 1676, when the Dutchman, Antonie van Leeuwenhoek, first observed living

organisms smaller than the single-celled animals (protozoa) he had seen in lake water. For more than fifty years, Leeuwenhoek wrote to the Royal Society in London about his discoveries, describing the microscopic organisms as 'mightily diverting', in the way they 'pulled their bodies and tails together, continuing their gentle motion'. A few years later, Herman Boerhaave, who was born in Leiden in 1668, argued that smallpox was caused by an infectious agent, not invisible poisons found in swampy air. Boerhaave is now considered the founder of clinical teaching and the modern teaching hospital but in his assertions about smallpox he was ahead of his time.

Limited progress was made until the 1870s, when the development of the compound microscope and high-quality lenses revealed living agents associated with numerous diseases, including typhoid, cholera, leprosy, tuberculosis and diphtheria. Scientists could now see the infectious microbes they worked with and they immediately tried to develop vaccines. Ferdinand J. Cohn published an early classification of bacteria in 1875, using the word 'bacillus' for the first time. The following year, working under Cohn's auspices, Robert Koch, who would become Germany's leading bacteriologist, published a paper on his work with anthrax, which identified a bacterium as the cause of the disease.

The first method for isolating a pure culture of bacteria was developed two years later, in 1878, by Joseph Lister (who later applied Pasteur's germ theory and became known as the father of antiseptic surgery) when he demonstrated the specific cause for the lactic fermentation of milk. This was the beginning of a flurry of microbiological discoveries; among the most well-known of which is the work of the French chemist, Louis Pasteur.

Born in the Jura region of eastern France, Pasteur was the son of a tanner, who grew up overlooking the stinking tanning pits.

After gaining a doctorate from the École Normale in Paris, Pasteur was eventually appointed professor of chemistry at the University of Lille. Here, his academic research was intended to benefit local industries including vinegar production and wine-making. Pasteur famously proved that liquids such as beer and milk went off because of the rapid multiplication of micro-organisms within them, and that heating the liquids to boiling point killed the micro-organisms. He went on to develop the technique of 'pasteurization' for preserving liquids.

Pasteur pursued his germ theory, applying his observations to solid matter. He won a prize from the French Academy of Sciences for demonstrating that the decay of meat was also caused by microbes. Based on his theory, Pasteur believed that disease in the body could be caused by a similar multiplication of microbes, or 'germs'. Pasteur studied chicken cholera in his laboratory, injecting chickens with the live bacteria and recording the fatal progression of the illness. His discovery that chickens were immune to cholera germs that had been left exposed to the air was made by accident: an assistant forgot to inject the chickens with a fresh batch of culture before he left for a month's holiday. When he completed the assignment on his return, the chickens developed a mild, rather than lethal, form of the disease. Once the birds were restored to full health, Pasteur injected them again with fresh bacteria. This time the chickens did not become ill at all. Pasteur reasoned that the factor that made the bacteria less deadly was exposure to oxygen. Unwittingly, Pasteur had developed a method of 'attenuating' a vaccine, making it live enough to generate immunity but weak enough not to cause the disease.

Five years later, in 1885, Pasteur applied his attenuated vaccine theory to rabies, growing the virus in rabbits then drying the affected nerve tissue, to weaken the virus. On 6 July 1885 he administered the vaccine to Joseph Meister, a nine-year-old boy who had been bitten by a rabid dog. Attenuated vaccines were a new, and risky, procedure and Pasteur's treatment was controversial in several respects: not only had he not successfully used the vaccine before in a human being (his previous attempts had been on a sixty-year-old man who left hospital after only one injection and a young girl who died before the second injection could be given) but he was not a medical doctor, and legally should not have been administering such treatment.

Pasteur was certain, however, that his inactivated viral vaccine would work. Moreover, it was the only hope for Joseph Meister, who would certainly have died of the disease. The course of the injections was spread over thirteen days, with a stronger (less attenuated) version of the vaccine being given every day. It worked; Meister did not develop rabies. Pasteur had successfully developed the first vaccine since Jenner's, one hundred years before.

Pasteur's contribution to microbiology was complemented by the equally important findings of German scientist Robert Koch, who published a paper in 1882 identifying the tuberculosis bacillus. He also developed the first of what became known as 'magic bullets'; chemicals that could kill specific bacteria. Pasteur and Koch were great rivals and the conflict that accompanied their two schools of work reflected the bad feeling that existed between the two countries, caused by the Franco-Prussian war of the early 1870s, in which Koch had served.

His experience as a medical officer on the front line led Koch to carry out scientific experiments on anthrax, which was prevalent in local animals. Building on Pasteur's germ theory, Koch extracted the bacterium from a sheep that had died of anthrax, grew it, injected a mouse with it and repeated this for twenty generations. Finally, he proved, in 1876, that the bacteria *Bacillus anthracis* was the cause of anthrax. Koch improved his techniques, solidifying liquids with gelatine and agar to create a solid medium for growing bacteria (his assistant, Julius Richard Petri, developed the dish that still bears his name) and staining bacteria to make them more visible under the microscope.

These improvements allowed Koch to identify the bacterial cause of tuberculosis in 1882 and cholera in 1883, and other researchers to identify the bacterial causes of typhus, tetanus and the plague. While others might have seen the agents of these diseases before, it was Koch's methods that allowed scientists to prove beyond doubt that the agents were the causes of disease.

Working with the newly identified bacteria for diphtheria and tetanus led to the second breakthrough in microbiology: the discovery of the antibody. In late nineteenth-century Berlin, the Prussian military doctor Emil von Behring and the Japanese researcher, Kitasato Shibasaburo, discovered a mysterious substance in the blood of guinea pigs that had been injected with a toxin. The substance neutralized the toxin they had been injected with, but no others. And mixed with serum and injected in fresh guinea pigs, the substance was found to be harmless.

Working in collaboration with Paul Ehrlich, Behring developed his theory of the antibody (or antitoxin as he first called it), developing a diphtheria antibody that was injected into children, with immediate and successful results. The discovery of the antibody heralded the birth of immunology but the first few years saw the field divide in bitter debate. In his book *The End of Plagues*, John Rhodes noted:

> In the last decade of the nineteenth century
> news was breaking almost every month of
> new microorganisms, new diseases, new

The Prophet tagging.

mechanisms; in the fevered inflammatory world of immunology two passionate debates were raging.[3]

Was inflammation an abnormal and harmful manifestation of disease or was it a positive defensive response to infection? And what was the nature of these 'defenders': were they invisible antibodies in the blood or white blood cells detectable under the microscope?

The German physician Rudolph Virchcow started the cellular theory in 1859 when he noted a large amount of white cell material in the spleen of a former cook. This was followed in 1884 by the work of the Russian biologist Elie Metchnikoff, who noticed 'wandering' white cells clustering around thorns inserted into jellyfish, establishing the importance of white cells in defence against disease. Metchnikoff joined the Pasteur Institute in Paris, which was pitted against the German school inspired by Koch's work. With the discovery of the anti-diphtheria serum the work of the Germans appeared to be ascendant, something that was reinforced when Richard Pfeiffer showed that the cholera germ from immunized animals contained antibodies but not white cells.

The next age of cellular immunity dawned in the aftermath of World War Two, when the British scientist, Peter Medawar, noticed the inflammation of skin graft rejection contained lymphocytes, a particular kind of small white blood cell, leading to the knowledge that B lymphocytes have the job of making antibodies, while T lymphocytes (T-cells) have the job of latching on to, and killing, virus-infected cells.

In *The End of Plagues*, John Rhodes concluded:

Despite the lack of knowledge about how vaccines worked in the 1890s, three kinds of vaccine had been discovered. Edward Jenner had pioneered the use of a related virus against smallpox; Louis Pasteur had used weakened pathogens to protect against cholera and rabies; and he had also discovered a chemically altered pathogen to protect against anthrax. In addition Emil von Behring had used an antibody to protect against diphtheria.[4]

The first viruses were not defined until 1892, when the Russian biologist Dimitri Ivanovsky discovered that a disease affecting tobacco plants could pass through porcelain filters, while bacteria could not. This was substantiated by the work of the Dutch microbiologist, Martinus Beijerinck, in 1898, when he filtered a solution carrying an agent that he called a 'virus', and by the discovery in 1906, by the Italian Adelchi Negri that smallpox was a filterable agent. Virus particles could not actually be seen, however, until the invention of the electron microscope in the 1920s.

The discoveries made at the turn of the twentieth century laid the foundation for a generation of enthusiastic public health officers to strike out in the new direction of combatting infectious disease. The scientists of the microbiological revolution had identified bacteria, understood the antibody, realized that viruses existed, even if they could not be seen, and grasped the crucial role that parasites played in certain diseases, identifying the protozoa (single-celled organisms with animal-like behaviour) that caused malaria and sleeping sickness.

In the closing years of the nineteenth century, a British doctor, Ronald Ross, and an Italian malarialist, Grassi, showed that malaria was transmitted to humans by the bite of the female mosquito belonging to the genus *anopheline*. This was followed by the findings of the Reed Commission in 1900, which discovered that yellow fever was transmitted to humans by the *Aedes aegypti* mosquito. This set the first significant global disease eradication effort in public health in motion.

'The most magnificent failure in public health'

The microbiology revolution was not just a journey of scientific discovery. It also shaped the language of the debate, setting out some of the possibilities and parameters for how public health officials could attack diseases. Words like 'invade', 'defeat' and 'combat' gave every health effort the thrust of a military strike, a battle won and a territory conquered; popular concepts in 1900 but ones for which enthusiasm waned as the century wore on.

Public health campaigns could have chosen very different tactics, such as improving the condition of people's lives, their accommodation, sewage systems or water supplies, all of which would have dramatically improved public health. But by focusing on specific bacteria and the insects that carried diseases, doctors and scientists were closing in on the causes of disease. Moreover, they became convinced that they could attack these diseases within a relatively short time, in a highly targeted way, uniformly across the world, without addressing any of the underlying social or economic problems that individual countries might face. The historian, Nancy Leys Stepan, noted: 'Whatever the disease in question was found, it was presumed to have the same cause and be open to elimination by the same methods, regardless of differences in the class, economic and geographical situation of the human populations involved. It was in this sense that eradication became international'.[5]

This theory was tested to the extreme in the effort to eradicate yellow fever, which began promisingly, with one of the most heralded discoveries since the birth of microbiology: the proof, given by US physicians in Havana in 1900, that the disease was transmitted by mosquitoes. Eventually, however, even the strongest proponent of the eradication campaign, Fred Soper, had to agree that it was 'The most magnificent failure in public health history.'[6]

Yellow fever, one of the first diseases proposed for eradication, was the subject of one of the longest efforts in history. If the complete wiping out of a disease were the only measure of success, it ended in failure, but the campaign exemplified some of the features that would distinguish future eradication campaigns: colonial 'imperial' origins, a sometimes unwarranted faith in medical knowledge and techniques, and a top-down authoritarian approach to running the campaign.

Much of the basis for running an eradication campaign was established by one man, Major William C. Gorgas, and vigorously prosecuted across the globe by another, Fred Lowe Soper. Judged by their own standards – absolute eradication and zero cases – the campaigns they waged against yellow fever, malaria and yaws were unsuccessful, yet in the process of their work they saved millions of lives.

Like polio, yellow fever was mysterious, with no known cause. It was also often deadly. Sometimes the disease seemed to invade communities from the outside, at other times it seemed local. Like polio, some people suffered only mild headaches and fever, while others (about fifteen to twenty per cent) became severely ill: turning yellow with severe jaundice, suffering from extremely high temperatures and experiencing internal haemorrhaging that resulted in bouts of horrific black vomit. Such terrible, terrifying, symptoms and unknown origins meant that, like polio, yellow fever took a grip on the public imagination out of all proportion to the likelihood of contracting the disease.

The horrors of this memorable affliction
were extensive and heart rending . . . In private families the parents, the children, the
domestics lingered and died, frequently
without assistance. The wealthy soon fled;
the fearless or indifferent remained from
choice, the poor from necessity. The inhabi-
tants were reduced thus to one-half their
number, yet the malignant action of the
disease increased, so that those who were
in health one day were buried the next.
The burning fever occasioned paroxysms
of rage which drove the patient naked
from his bed to the street and in some
instances to the river, where he was drowned.
Insanity was often the last stage of its
horrors.

Samuel Breck's account of a yellow fever
outbreak in Philadelphia, 1793[7]

Yellow fever was a disease of Africa and the Americas that reached
the USA as a result of the slave trade. After the Civil War, in 1865,
yellow fever epidemics moved south, affecting Mississippi, Florida
and Texas, with the last major epidemic, in 1878, claiming twenty
thousand lives in the Mississippi valley. Even with a decline in
cases, however, US authorities dreaded the disease, causing the
establishment of the first national public health organization, the
American Public Health Association and the passing of the 1878
Quarantine Act. Since the disease often seemed to be brought in
from outside, increased trade and shipping were thought to pose a
particular threat.

The 1898 American occupation of Cuba, which until then had been in the final stages of throwing off Spanish colonial rule, led to a particular focus on the growth of yellow fever cases in Havana and in the US military. During the US military blockade, Havana had remained relatively free of yellow fever but the reopening of the port in August 1899 led to an influx of immigrants and sailors. By the end of the year Havana was in the grip of a yellow fever epidemic, which caused 1,400 cases in 1900. Fearing an epidemic among American troops and import into the USA, in May 1900, the authorities set up the US Army Yellow Fever Commission, known as the Reed Board. The Reed Board consisted of four members: Dr Walter Reed, Dr James Carroll, Dr Jesse Lazear and Dr Aristides Agramonte. None of the members was an expert in yellow fever but together they set up a series of experiments at Camp Colombia, some miles outside Havana, to test hypotheses about the disease. One hypothesis was proposed by a Cuban doctor, Carlos Finlay, who had carried out inconclusive inoculation experiments with mosquitoes and who now demonstrated that the most likely (of the more than six hundred Cuban species) mosquito involved in yellow fever was *Aedes aegypti*, a 'beautiful insect marked by artfully arranged black and white lines and spots'.[8]

Based on these conclusions, the Reed Board began a series of human experiments, conducted by Jesse Lazear, with infected mosquitoes but in the second round all three subjects, and Lazear, fell ill with yellow fever and died. Although it was an ill-fated conclusion, Walter Reed announced at the American Public Health Association that the experiments had shown that *Aedes aegypti* transmitted yellow fever.

The announcement contained a key omission: the Reed Board had not attended to another paper made available by Dr Henry R. Carter, a quarantine expert with the US Marine Hospital Service.

Carter informed them that, based on an investigation he had conducted on yellow fever in Mississippi, there was a two-week interval between the first and secondary cases of yellow fever, which indicated that the 'germ' underwent some sort of change in its environment. If the Board had grasped this they would have understood why the bite of the mosquitoes infected with yellow fever did not transmit the disease in the first round but did a few weeks later, with deadly effect.

Once the role of *Aedes aegypti* had been established, the Reed Board promptly moved into campaign mode. As no drug was then available for treating yellow fever, the best course of action seemed to be to wipe out the carriers of the infection, the mosquitoes. The assumption that *Aedes aegypti* was the *only* type of mosquito to carry yellow fever formed the scientific bedrock of the elimination effort. The governor general of Cuba, William C. Gorgas, initiated an eight-month campaign in Havana to 'set the pattern of attempts to control yellow fever for the next thirty years'.[9]

Although he was a physician, Gorgas was in spirit a military rather than a medical man. He set about his plan to rid Havana of mosquitoes with precision, planning and thoroughness. Although he was initially far from convinced by the findings of the Reed Board, he followed orders to concentrate his sanitation efforts on anti-yellow fever mosquito elimination. This involved switching two-thirds of the Sanitary Department's crews from cleaning streets on to a house-to-house campaign that used chemicals and oil to destroy mosquitoes and their breeding grounds. Houses were fumigated (no easy task when the act of burning chemicals inside poorly constructed shacks often burned down the buildings themselves), drinking tanks were sealed and open surfaces of water were covered in a thin layer of oil to smother the mosquito larvae. In addition, Gorgas asked for the authority to station inspectors on the railroads into the city and stop non-immune people from entering unless they had been in quarantine for a week.

The campaign against yellow fever was run as a military occupation under martial law, which made matters easier. The approach was technical and medical from the very beginning. It had four strands that were applied to all subsequent eradication campaigns. First, Gorgas 'noted with satisfaction that in Havana he had got rid of yellow fever without tackling any of the traditional problems of public health'.[10] At no stage were environmental measures, such as installing a piped water supply or replacing shacks with better housing, ever considered. Second, the yellow fever campaign was run with a singular focus on one disease and operated outside the existing structures of public health, using mosquito squads removed from other sanitation duties. The US wanted immediate, specific results and was not interested in improving Cubans' general health, nor much concerned that people might be suffering from more than one disease: 'The specialized service was to be run like a unit of the army, with teams or squads of mosquito killers working under strict hierarchical discipline.'[11]

Third, Gorgas introduced precise mapping and recording of data, demanding that those working on the mosquito squads followed a precise route, fining them if they deviated. Havana was mapped for mosquito breeding locations, dividing the areas into inspection districts and marking every house. Each inspector was accompanied by a squad of 'oilers' who poured oil on to all puddles of water. The inspector was required to fill in detailed forms and submit them every night. The data were then consolidated into an overview monitored by the sanitation department. Anything where mosquitoes might breed – old bottles, discarded cans, any container-like rubbish found lying in yards and gardens – was inventoried and removed. House gutters were cleaned and straightened to prevent water building up; screens were put in windows wherever possible.[12] In total, twenty-six thousand possible breeding grounds for mosquitoes were identified. Fourth, as a military man, Gorgas demanded and received complete authority over how the campaign was run, bringing in special ordinances to ensure the public complied with his orders.

While the population of Havana appeared largely indifferent to his efforts, the results were remarkably successful, with the city reporting only a few cases of yellow fever in the high season of 1901 and none at all in 1902. 'This is evidence of the practical demonstration of mosquito theory,' Gorgas concluded. The US authorities agreed with him, soon planting his mosquito extermination campaign, as a kind of 'surgical intervention' in other parts of Cuba, in Latin America and Brazil and, famously, in Panama.

In 1905, with what was viewed as admirable efficiency, New Orleans stamped out a yellow fever epidemic that claimed 452 lives. A similar campaign between 1903 and 1906 rid Rio de Janeiro of yellow fever for twenty-eight years. But the greatest significance should perhaps go to the anti-mosquito campaign waged in Panama in 1904, where Gorgas wrested sufficient control from the Panama Canal Commission to stop a 1905 epidemic among workers building the Panama Canal. The disease did not return until 1954.

Such was the high incidence of disease (yellow fever and malaria) that without Gorgas's anti-mosquito campaign it is doubtful that the canal could have ever been completed. The cost was complete authoritarian control, a price tag of two million dollars a year and the enormous use of chemicals; more than 189,270 litres of paraffin oil and 140kg of sulphur a month for house fumigation.

Gorgas proved that under military rule, and at enormous environmental cost, mosquitoes could almost be wiped out in a small slice of a foreign country. The US authorities and others noted that their representatives, in civilian or military form, could occupy parts of the tropics more safely and workforces could kept healthy enough to conduct large work projects without any kind of intervention in the social or health system of the country concerned. As Gorgas himself put it, 'Life in the tropic will be more healthful for the Anglo-Saxon than in the temperate zone.'

Gorgas died in London on 3 July 1920, while on his way to spread his anti-mosquito campaign to West Africa. He not only received an honorary knighthood from King George V, presented at his deathbed, but also a special funeral at St Paul's Cathedral. In the months before his death, however, he had passed the baton to someone would turn his campaign into a crusade: Fred Soper.

'The General Patton of entomology'

Fred Lowe Soper, a native of Kansas, was a young doctor recently recruited to the International Health Division of the Rockefeller Foundation when he heard Gorgas lecture at the Johns Hopkins University School of Hygiene and Public Health in Baltimore. Hearing the lecture, Soper feared that he had 'arrived on the scene too late' to be part of the 'glorious' eradication of yellow fever, as Gorgas made everything seem so 'simple and straightforward'. Soper's life work would prove that he could not have been more wrong. But not for want of trying.

Both Gorgas and Dr Henry Carter had floated the idea that yellow fever could be entirely wiped out, with Carter stating that for 'yellow fever the sanitarian should be satisfied with nothing short of elimination. It is easy.'[13] Soper perhaps did not believe that it was 'easy' but he certainly believed that it was possible and that eradication via wiping out mosquitoes was the only means to achieve his goal, even when other alternatives, such as a yellow fever vaccine, were developed. 'Fred Soper was the General Patton of entomology', Malcolm Gladwell noted in the *New Yorker* in 2001. 'His special contribution was to raise the killing of mosquitoes to an art . . . His method was to apply motivation, discipline organization and zeal, in understanding human nature.'[14]

Soper was not particularly motivated by humanitarian purpose. In later life, some of his American colleagues refused to consider him for jobs based in America, on the basis that his approach to public health was 'fascist'. We do not know why he chose to work for the International Health Division (IHD) at the Rockefeller Foundation in 1919. The IHD, founded six years earlier, was at the time deeply engrossed in its first (failed) disease eradication effort against hookworm. At Soper's interview, the head of the IHD, Wycliffe Rose, mentioned that the foundation was also working on yellow fever, which they had committed to eradicating in 1915, estimating that it would take five to ten years. Even before Soper's arrival yellow fever eradication absorbed most of the Rockefeller Foundation's funding for the Americas, accounting for half of the budget between 1925 and the mid-1930s.

Soper began work for the IHD in 1920 and was sent on his first posting to Brazil to work on hookworm control. He quickly learned that disease control was not as simple as sometimes suggested. People failed to turn up for follow-up treatments, the size and difficulty of the terrain was breathtaking and installing the necessary latrines was not always possible. When Soper returned to Johns Hopkins University for a year of further study in 1922 he was shocked to discover that many doubted whether hookworm eradication was indeed possible. Nonetheless, he was posted to Paraguay for four years, where he was charged with setting up a hookworm eradication programme, and where he learned many of the skills he would deploy in the rest of his career, including mapping, surveillance and managing his own team.

Despite strenuous efforts by IHD officers, hookworm eradication was doomed to failure. From the beginning, the Rockefeller Foundation had adopted the kind of biomedical approach to disease control that would characterize later eradication efforts. Lack of US government funding or a centralized approach to healthcare left a gaping hole for philanthropy to step in to. Without nationalized healthcare, the US had only science, not society, to fall back on. Or, as Edmund Ramsden puts it in the *Oxford Handbook of the History of Medicine*: 'Big science *is* America's solution to sickness.' Big science attempted to curb hookworm without investment in improving people's lives, such as installing latrines, providing clean water or buying shoes. These actions would have helped the hookworm effort immensely but the Rockefeller Foundation was unwilling to do any of them, insisting that they were the responsibility of the countries themselves. By the end of the 1920s the IHD had quietly admitted that two of its major programmes, the hookworm eradication effort and the tuberculosis programme in France, were failures. Soper found himself transferred to working on yellow fever eradication. Ironically, he was about to become immersed in yellow fever eradication at the same time that his employers at the Rockefeller Foundation began to believe that it was another impossible mission.

As the biggest and most well-funded American foundation working in health, the focus of the Rockefeller Foundation on yellow fever gave the disease a prominence historians have subsequently claimed it did not deserve. Much the same criticism would later be levelled against polio eradication. Founded by the Standard Oil owner, John D. Rockefeller, as an attempt to improve the family's image after a damning book on the company was published in 1904, the Rockefeller Foundation established the IHD to reform medical education in the USA and spread public health work abroad, eventually setting up public health schools in twenty-one countries and entrenching American ideas about eradication across the world.

More air travel and shipping meant that more people and goods were moving around the world – making the threat of global diseases seem more acute – and stopping the disease at its source appeared sensible. In the countries themselves, however, so much funding for yellow fever seemed almost perverse in the face of so many more pressing issues. 'The IHD directed resources at yellow fever out of all proportion to the disease's impact on local health', wrote Heather Bell. 'Local resistance to the global goal – whether by the general population of a government – was regarded by the IHD and its allies as selfish and irresponsible.'[15]

In 1918, the IHD again rejected improving sanitation and housing conditions in favour of Gorgas's methods in its first yellow fever campaign in Ecuador. Those methods appeared to be successful and, after clearing the city of Guayaquil of yellow fever in six months, the IHD moved its campaigns into Mexico and Peru. A successful move into Brazil was heralded by the IHD as the beginning of the end for yellow fever in the Western Hemisphere. Such optimism turned out to be completely unfounded, however, when an outbreak in Rio de Janeiro, and forty other towns in the state in 1928–29 proved that the assumptions that underpinned IHD's work were wrong.

It had been – incorrectly – assumed that yellow fever was an urban disease with no animal host, only transmitted by the *Aedes aegypti* mosquito, although many Latin American doctors knew that the disease was often found in remote rural locations and this strain sometimes caused a mild form of the disease in humans. By the early 1930s research results demonstrated that there was indeed a 'jungle yellow fever' in which the primary hosts of the virus were monkeys, not humans and that, therefore, yellow fever could never be eradicated without killing off all forest animals, something deemed impractical and unacceptable even in those days.

The startling discovery that yellow fever could never be eradicated coincided with the IHD's realization that it had spent years investing in a failed yellow fever vaccination project conducted by the flamboyant and eccentric bacteriologist Dr Hideyo Noguchi, who spent ten years conducting a long and confusing series of experiments on mammals and birds, uncorroborated by other scientists. Eventually, after much external pressure, a Rockefeller Foundation Yellow Fever Commission travelled to Nigeria and examined and undermined Noguchi's work but not before Noguchi himself died of yellow fever, his faith in his work undimmed. Usefully, the Yellow Fever Commission discovered, in the process, a strain of yellow fever that formed the basis of Max Theiler's successful 17D yellow fever vaccine in 1937.

By then it was clear that yellow fever eradication was impossible, and that vaccination was a suitable alternative to mosquito elimination, but Fred Soper remained strangely undeterred. Soper had been posted to Brazil in 1927 and by 1930 was head of the country's yellow fever eradication programme, running the equivalent of a large government department focused on only one disease. Without the knowledge of 'jungle yellow fever' that was soon to come, Soper believed that the 1928 Rio de Janeiro epidemic was caused by an attempt to merge the yellow fever programme into a general health service. He remained committed to the concept of single-disease campaigns for the rest of his life.

The incoming authoritarian president, Getulio Vargas, was shocked by the re-emergence of a disease that he believed undermined the notion that Brazil was a civilized country attractive to European immigrants. Consequently, his government agreed to allow the Rockefeller Foundation to expand its work into all areas of the country and accorded a joint commission great power. Soper took charge of the Rockefeller Foundation–Brazilian Co-operative Yellow Fever Service in June 1930. He ran it for twelve years, becoming what the malaria expert Paul Russell termed, 'The prophet of disease eradication'[16] who developed a 'Bible' for the ultimate eradication campaign, including the most specific record-keeping ever undertaken. In his personal notes Soper included a quotation from a veteran of the hookworm eradication campaign: 'Experience proved that the best way to popularize a movement so foreign to the customs of the people . . . was to prosecute it as though it were the only thing in the universe left undone.'

In his work he followed 'perfectionism to the end of the line'. If Gorgas's mantra was that to kill a mosquito it was necessary to think like a mosquito, Soper believed that: 'In order to kill mosquitoes it was necessary to think like a mosquito inspector.' This he did, for in his heart he was an administrator, commanding teams of inspectors reporting directly to him in all they did. Soper lost 12kg in weight through his dedication in following his inspectors about as they worked, ensuring that he understood exactly what they did and could exert fuller control over them. Legislation passed by the president himself gave Soper a free hand in hiring and firing employees and imposing whatever sanitary measures he thought fit. Soper paid his inspectors well but the work was exacting. He spent a quarter of his budget on checking up on the inspectors; famously, when one of the inspectors avoided being killed in an explosion because he had chosen that day not to go to work, Soper fired him for dereliction of duty.

'He was very cold and very formal,' the tropical disease expert Andrew Spielman told Malcolm Gladwell. 'He always wore a suit and tie. With that thin little mustache and big long upper lip, he scared the hell out of me.'[17]

In addition to striking fear into his employees, Soper upset local people by demanding post-mortem samples from dead relatives. Five health workers were killed carrying out his orders but Soper pressed on regardless. If his work did not make him popular, it was nonetheless effective. His innovation of sending in follow-up 'mother squads' of inspectors a few days after an initial inspection meant that many homes could be certified free from infection for the first time.

In 1933 Soper discovered that the *Aedes* mosquito had been wiped out in the coastal cities of north-east Brazil and he began to promote the idea that only wiping out the species completely would achieve eradication. Soper's thinking followed that of Louis Pasteur who had said, 'It is within the power of man to rid himself of every parasitic disease.' Soper believed that not only was it within his power but that it was his absolute duty. 'This is the fundamental difference between those of us in public health who have an epidemiological perspective and people, like Soper, with more of a medical approach,' Andrew Spielman said. 'We deal with populations over time, populations of individuals. They deal with individuals at a moment in time. Their best outcome is total elimination of the condition in the shortest possible period. Our first goal is to cause no outbreaks, no epidemics, to manage, to contain the infection.'[18]

By the mid-1930s Soper faced several enormous obstacles to his mission. As he reported back to the Rockefeller Foundation, the existence of jungle yellow fever and discovery that monkeys were the animal hosts of the disease made the goal of eradication impossible. In addition, the new yellow fever vaccine appeared to challenge his assertion that only killing mosquitoes would eliminate the disease. Soper changed his emphasis from complete eradication to complete eradication of urban yellow fever and maintained that a vaccine programme, which he had been involved in successfully administering in Brazil, in no way contradicted the need for mosquito control.

Nonetheless, he no longer commanded the support of the Rockefeller Foundation and in 1935 the new director of the IHD, Dr Wilbur A. Sawyer, wrote to complain that Soper was now effectively the head of a government department, which was 'hardly consistent with our general policies'.[19]

'Suggestion to redefine eradication is not acceptable. . .'
How should history judge Fred Soper? A racist? A colonialist of the highest order? A life-saver? Or as a rather cold-blooded and single-minded idealist whose actions and interactions with the world, like those of the Rockefeller Foundation itself, were rather more nuanced than polemicists admit? Soper admitted his 'capacity for fanaticism' had undermined his role at the IHD, but his experience working on malaria in World War Two strengthened his belief in disease eradication. He found that his way of thinking came back into fashion in the post-war years, when the newly created World Health Organization in Geneva committed to a global malaria eradication programme, of which he was a key proponent.

In 1947 Soper took over the directorship of the Pan-American Health Organization (PAHO) and used all his powers and efforts to turn it from a small organization of thirty-two staff with a tiny budget into a large public health body with a vastly increased budget and a staff of 750. In the process he brokered an agreement to have PAHO incorporated as the regional office for the WHO and persuaded the organization to lead continent-wide eradication efforts against yellow fever, yaws, smallpox and malaria. He also mentioned getting rid of polio, cholera, leprosy, rabies, the plague and tuberculosis.

Soper's goals (which were not at all political) fitted very neatly with the aims of the USA, which supported the idea of technical interventions to stop diseases that would free up more labour for economic advancement and global trade but would never support any public health effort that could be construed as 'socialized medicine'.

'Fred Soper with modest ambitions would not be Fred Soper,' Malcolm Gladwell noted, going on to add that Soper's greatest achievements arose from his absolutism and his determination to save as many lives in as short a period as possible. 'There are few people in history to whom so many owe their lives', Gladwell concluded.[20] Soper planned to rid the world of disease and in two instances, yellow fever and yaws, he was nearly successful. But nearly was not near enough.

Yaws is a bacterial infection spread by skin contact between people living in unsanitary conditions. Soper's campaign reduced the disease by ninety-five per cent in the 1950s and 1960s but as attention dropped and the sense of urgency passed, a small but crucial number of latent cases was missed and by the end of the 1970s yaws was as prevalent as it had been before the beginning of his efforts. Attempts to integrate yaws work into the existing health structures of countries was deemed a failure, as many were incapable of meeting the demand. These would be crucial lessons for the polio and smallpox campaigns.

In yellow fever, Soper saw his legacy even more painfully dismantled. As head of PAHO, Soper devoted the resources of the organization across the continent to his mission to eradicate the mosquito. By the mid-1950s maps of the region showed that nearly all yellow fever cases were in remote rural areas. Nearly, but not quite all. Soper kept at it, year after year and decade after decade, demanding that every country in the region was unrelenting in its efforts. But the truth was that with its vast costs, support for eradication was waning, not least in the US, which believed the yellow fever vaccine was more than adequate for its needs. Soper's replacement as the head of PAHO in 1958, Dr Abraham Horowitz, maintained the eradication programme but shifted the organization towards a focus on better public health, clean water and better sewage systems. A decade later, PAHO delegates, including the US, were openly resistant, questioning the cost-benefit ratio of keeping the eradication programme; so expensive and so few cases.

The yellow fever eradication programme was finally abandoned in 1970 and, inevitably, as Fred Soper predicted, the *Aedes aegpti* mosquito returned, bringing with it unanticipated viruses, such as Dengue fever and Zika. The most recent yellow fever outbreak, which began in Angola in December 2015 and spread across the Democratic Republic of Congo, has so far involved a suspected six thousand cases (according to WHO) and claimed six hundred lives. Today the numbers of mosquitoes are now greater than ever in urban areas. Soper's methods were autocratic, insensitive to the communities they served and disastrous to the environment, but they were nonetheless effective methods of disease control.

Fred Soper lived until 1977 but never lost sight of the end-game. 'Suggestion to redefine "eradication" is not acceptable,' he wrote in his notes for his final speech to PAHO as its departing chief. '*Must have the courage of their convictions.*'

'Let us spray' became the mantra of public health officials during the 1950s as DDT spraying became the global answer to mosquito control and disease eradication, until Rachel Carson's book *Silent Spring* drew attention to the dangers the chemical posed.

6

THE MOSQUITO HOUSE

Every time malaria has been brought under control in history, manage-
ment of the mosquitoes has been the central approach.
Dr Peter Agre, Director of the Johns
Hopkins Malaria Research Institute[1]

In an insectary in the city of Baltimore the mosquitoes are buzz-
ing. Harvested eggs are allowed to hatch in petri dishes and
enclosures. When the larvae mature into mosquitoes they fly
around in nets, 'kind of like a cheesecloth-enclosed box' 20cm in
size, where the temperature and humidity are closely monitored
by scientists. 'Mosquitoes seem to be very fussy about what they
prefer,' mused Dr Peter Agre, director of the Johns Hopkins Malaria
Research Institute.[2] Agre's facility at the institute studies innovative ways of combatting malaria, the parasitic disease which has
thus far adapted to every human effort to combat it and, in the
process, set back the cause of disease eradication for decades.

The work of the Johns Hopkins team includes developing a
transmission-blocking vaccine, repurposing known medicines to
treat malaria, and studying the cellular and molecular events that

allow the parasite to live and replicate in the mosquito. However, their recent work to develop a genetically modified *Anopheles* mosquito has been the focus of much attention. This mosquito is highly resistant to the malaria parasite and passes on this trait to its offspring when they are eventually released to mate in the wild. The modification uses a new gene-editing technology, CRISPR (Clustered Regularly Interspaced Short Palindromic Repeats), which enables a gene on one of the two chromosomes inherited by a mosquito to copy itself to the other, making sure that all the off-spring inherit it. Using this technique, malaria-resistant genes should spread much more quickly through mosquito populations. According to Peter Agre, it is a process of 'building a better mosquito', rather than killing all mosquitoes, as Fred Soper fanatically tried to do.

'The notion that mosquitoes could be genetically modified to resist the malaria infection has been enforced by several investigators,' Agre said. 'And it's not really to make them better mosquitoes; it's to make them better at resisting malaria. When they transmit malaria, the mosquitoes themselves are weakened. They're in a diminished state. But they're still able to pass on the infection from one human to another.'

Malaria-resistant mosquitoes should be stronger in the wild, Agre argues. 'They'd be fitter than the usual mosquitoes. And the humans would have a great advantage of not getting malaria,' although he added 'that doesn't mean the mosquitoes wouldn't bite people'.

Agre received the 2003 Nobel Prize for chemistry for discovering aquaporin water channels, the 'plumbing system of cells', before turning his attention to malaria. As well as directing the programme in Baltimore, he oversees Johns Hopkins's fieldwork in Zimbabwe and Zambia, where the team has constructed a massive

'Mosquito House', a semi-sealed testing ground for the next stage of the genetically modified mosquitoes project. 'Mosquitoes seem great in these small caged enclosures here in a temperature-controlled humidified room but that doesn't mean that this approach will necessarily work in the wild,' Agre said. 'In the Mosquito House in Zambia we can see how the mosquitoes do compared to wild mosquitoes without letting them go into the countryside.'

The modified mosquitoes are 'marked' so researchers can compare them to wild mosquitoes, because the next, and probably more important question in terms of relevance to malaria, is to see if the modified mosquitoes can transmit the malaria parasite. This will require a new set of permissions, Agre said: 'No African government is going to give somebody free rein to raise malaria-infected mosquitoes if there's some chance that they could escape and infect the population.'[3]

Agre's team are not alone in their innovation. The once-dormant field of malaria research is teeming with activity and money. The logistical challenge of delivering an effective vaccine to remote regions faced by the polio and smallpox eradication campaigns was not an easy task but it seems simple and straightforward compared to malaria. The complexity of malaria means that dozens of research teams, a disparate coalition of different groups, operating in different areas, with no coordination, are working on every aspect of controlling and eliminating the disease. 'The malaria world is vast,' said Hugh Sturrock, an epidemiologist who works for the Malaria Elimination Initiative at the Global Health Group at the University of California, San Francisco, under the leadership of Sir Richard Feachem. Malaria researchers are more likely to work across diseases, such as in Sturrock's area of epidemiological mapping, than in a vertical single disease campaign structure. Somehow, their innovations must be pulled together into a coherent whole.

The 'E' word again

The collapse of the malaria eradication campaign of the 1950s and 1960s made talk of elimination or eradication unthinkable. As Sturrock put it, any muttering of 'the E word' resulted in the person 'being sort of sent out of the room'.[4] Although activity resumed in about 2000, talk of a wider plan, beyond disease control, seemed premature. But once again, Bill Gates blasted away caution and set out a bold new agenda. In October 2007, the Bill and Melinda Gates Foundation confounded the global health community by announcing that the goal of the foundation was to achieve, over several decades, the complete eradication of the disease. The goal included the development of new anti-malaria drugs, a malaria vaccine and other scientific innovations. Much to the surprise of those who believed malaria eradication was off the table forever, Dr Margaret Chan, director general of the WHO (the organization so inexorably involved in the failures of the previous eradication campaign), supported the idea.

Malaria kills nearly half a million people every year but the malaria mortality rate has declined by half since 2000. The WHO estimates that malaria campaigns have saved the lives of 3.9 million children in Africa, where most malaria deaths occur. However, these figures are disputed by some, including the journalist Sonia Shah, as the WHO changed its method of classifying statistics in the middle of the decade, making the actual fall far less dramatic. Nonetheless, deaths from malaria could fall by a further ninety per cent in the next fifteen years, the WHO predicts. 'We can win the fight against malaria,' Dr Margaret Chan said. 'We have the right tools and our defences are working. But we still need to get those tools to a lot more people if we are to make these gains sustainable.'[5]

The prediction of the Gates Foundation that malaria can be eradicated completely by 2040 would signal a remarkable turnaround for a disease that was endemic in almost every country in the world at the beginning of the twentieth century and has proved a fearsomely adaptive opponent to all attempts to wipe it out.

Conquering 'the little brute'

Malaria starts when a person is bitten by an infected mosquito, which injects a number of protozoan parasites into the blood-stream. Unlike yellow fever, which is transmitted by one type of mosquito, more than forty species of mosquito carry malaria.

'Your scent tickles the mosquito's antennae. Your movement catches her eyes,' wrote Andrew Spielman, then 'the heat generated by your flailing muscles guides her to the most radiant spot on your body.'[6] There she (for it is always the female) bites – tastes blood – and fills her stomach, tripling her body weight. Finally, she pulls away and rests for up to forty-five minutes on the nearest upright surface, before flying off with enough blood inside her to allow her to start reproducing new offspring. She leaves behind an unsuspecting victim, infected with parasites that travel to the liver where they multiply rapidly, infecting the red blood cells. The victim will start to feel ill one to four weeks later, when flu-type symptoms indicate that the parasites have broken out of the blood cells. At this stage, other mosquitoes pick up the parasite when they bite an infected person, and pass it on again.

Of the four types of malarian protozoa that cause illness in humans (a fifth type infects primates but has also been known to cause human malaria), eradication efforts usually focus on two: *Plasmodium falciparum* and *Plasmodium vivax*. *P. falciparum* is the most deadly, accounting for almost all the 528,000 deaths from malaria in sub-Saharan Africa in 2013. *P. falciparum* is carried by the *Anopheles gambiae* mosquito, an 'efficient' carrier of the disease, which feeds only on humans, adding to the difficulty of controlling the spread. *Plasmodium vivax* is more widespread in other parts of the world, and less lethal, although it remains dormant in the liver and causes frequent relapses and ongoing illness and weakness in the victim. In some people, the parasite can lie dormant for long periods and up to eighty-five per cent of people infected with malaria show no symptoms at all.

Until the mid-twentieth century malaria was a disease that millions of people around the world simply accepted and learned to live – or die – with. Ancient Romans understood the connection between malaria and the wet, marshy breeding grounds of mosquitoes, trying to drain the Pontine marshes to make the area healthier, but not even the temple they built to the goddess of fever on the Palatine Hill could spare them. In ancient China, men travelling to malarial areas were told to arrange for their wives to remarry before their departure, while malaria was one of the factors that prevented Genghis Khan from invading Western Europe. In England, malaria plagued the low-lying areas of Kent and Essex four hundred years ago, stretching far up the banks of the Thames.

Malaria did not change geographically as a disease until the nineteenth century, when urbanization drew people from swampy agricultural areas into towns and cities. As this happened, and living conditions improved in Europe and North America, the disease shifted south, away from temperate Europe and into tropical areas, where colonial agricultural development opened up new breeding grounds for mosquitoes, as well as introducing them to huge numbers of labourers who had no immunity to the disease. In parts of the world such as Northern Europe, therefore, malaria receded without conscious human efforts to stop it but where it remained, it was tenacious.

As with smallpox, polio and yellow fever, the world of malaria research is filled with curious and eccentric characters. A French army doctor, Charles Laveran, was the first to see malaria parasites in human blood in 1880, but a British doctor, Ronald Ross, gets most of the credit for demonstrating how the malaria parasite enters the human bloodstream. Ross was considered an amateur and interloper, a man who transported himself from a life of leisure in the Indian Medical Service, where he played golf, wrote poetry and published a romantic novel, and turned into a pseudo-scientist, with a sudden determination to prove malaria's growth and development in mosquitoes. Fellow scientists, such as his arch-rival, the eminent zoologist Giovanni Grassi, fumed at Ross's bare-faced cheek. As Andrew Spielman noted, the goal Ross set himself was, 'A tall order for a man who hadn't been able to find the parasite on his own, didn't know one species of mosquito from another and had never dissected one.'[7] Nevertheless, Ross persisted and after years of conducting experiments on reluctant local people, who often thwarted his plans by running away, he was finally able to infect, kill and dissect a mosquito in 1897 and so spot a malaria protozoan, or 'the little brute', as he called it.

In subsequent decades scientists discovered which types of mosquito carry malaria and which do not, as well as identifying the different types of malaria parasite. None of this was of much help, however, for those who were unfortunate enough to contract malaria. There was no treatment other than quinine, a fever-quelling alkaloid made from tree bark. Quinine, unfortunately, had such a bitter taste and unpleasant side effects people often only took it under duress. An Italian experiment in the early 1900s to distribute free quinine failed when it was found that few took up the offer.

By the latter half of the nineteenth century malaria had spread across the globe. By the beginning of the global malaria elimination campaign in 1955, it was attacking an estimated 250 million people and killing 2.5 million. Epidemics greatly reduced people's ability to work, while death rates soared. The USA estimated that in 1938 malaria in the southern states cost the economy $500 million and in the years leading up to World War Two the disease was considered the single biggest health problem in India. One expert wrote: 'The problem of existence in very many parts of India is the problem of malaria.'[8]

While the problem of malaria was very well known, opinion was sharply divided on how to combat it. Experts roughly fell into two camps: those who wanted to kill mosquitoes and found ever-better chemicals to do so (the USA supported this approach), and those who wanted to treat the disease by improving people's living conditions and lives, draining swampy areas and tackling rural poverty; the Italian school of thought. While experts in this 'house divided' clashed for years, a middle way became clear when even mosquito eradication-focused experts, such as Lewis Hackett from the Rockefeller Foundation, came to appreciate that while chemicals and mosquito elimination were a vital tool, understanding human activities – such as building new irrigation ditches and giving people sufficient resources to put screens on their windows – also had a significant effect. By the 1930s malaria was understood as a complex disease, existing in many different forms and responding best to highly localized targeted control and treatment plans.

Perhaps nothing reinforced the burden of malaria more heavily than the enormous problem it presented to the Allies throughout World War Two. In the Pacific arena, tens of thousands of troops fell ill with the disease, rendering them unable to fight. In Italy, the Allied landings in Sicily were affected when thousands of soldiers, wearing regulation shorts, were bitten by mosquitoes and hospitalized with the disease; eight thousand men contracted malaria.

Further north, the retreating Nazi forces reversed the pumps that drained the Pontine marshes, reinfecting half the local population with the illness. In Egypt, Fred Soper's mosquito eradication 'bible' was implemented by officials after the devastating return of epidemic malaria in 1943 killed 180,000 Egyptians and threatened thousands of Allied troops in North Africa the following year. Soper's manual referred to the use of Paris Green, the chemical he had used so successfully in Latin America, but by the mid-1940s the Allies were already looking at another, soon-to-be rather infamous insecticide, developed in Switzerland in 1941.

'Let us spray'

Dichlorodophenyltrichloroethane (DDT) was first synthesized in the late nineteenth century but its properties as an insecticide were not appreciated until Paul Muller began tests for the Geigy Company in Switzerland. The USA and UK began testing DDT in 1942 and found that it was a highly effective insecticide, putting it to use to delouse the population of Naples during a typhus outbreak in 1943. DDT appeared far superior to other insecticides, which were often highly poisonous to humans and short-lasting. By contrast a small amount of DDT covered a large area, could be easily sprayed and had long-lasting effects. Most crucially, it appeared to be safe for humans, as it did not dissolve in water and could not be absorbed through the skin. Even food sprayed with DDT was deemed safe to eat.

Soon, widespread spraying of DDT was deployed against malaria. Spraying covered wide areas of Italy, where the Nazi actions had caused fresh epidemics. On the island of Sardinia, large areas were made habitable for the first time. Working with the malariologist Paul F. Russell, Fred Soper undertook a reconnaissance flight over the island in July 1945 and quickly concluded that 'from available information and what little I had seen it appeared that anopheles eradication in Sardinia might be entirely feasible if the materials, transportation, money and authority could be made available'.

The truth was that malaria cases had fallen in Italy during the 1930s, as a result of the Mussolini government's policies of land rec-lamation, drainage and improved standards of living. Nonetheless, malaria remained endemic on Sardinia but the logistical obstacles faced by an eradication campaign were severe. The island was large, with massive mountain ranges and deep ravines and a popu-lation that lived almost entirely in villages. Three species of malaria-carrying mosquito were identified, and more than a mil-lion breeding grounds.

It took three years, millions of dollars and sixty-five thousand workers to spray 267 tonnes of DDT over the island. The massive effort stopped the transmission of malaria but crucially failed to eradicate its vector (the agent that carries the disease): the mosquito lived on. The Sardinia campaign had shown that it was possible to use massive quantities of DDT, sprayed over a short period of time, to halt transmission but the window for success was very small, as mosquitoes developed resistance alarmingly quickly.

This lesson largely went unheeded. Confidence in DDT was so great that in 1947 Fred Soper, in his role as the newly elected head of PAHO, flew to Mexico City to announce that he was 'forced into predicting the end of malaria in the world during the next ten to fifteen years'.[9] Seven years later, the World Health Assembly endorsed the plan proposed by the Americas region for the total eradication of the disease and launched the Malaria Eradication Programme (MEP), the consequences of which weighed heavily on the global health community for decades.

Unlike the complex and localized understanding of malaria that had developed in the 1930s, the MEP supported eradication through massive widespread DDT spraying, following the same methods in every country where the programme existed. As predicted, insect populations crashed, as trucks wafting DDT fumes, and men with packs on their backs and spray guns in their hands cut a swathe across Greece, Italy and South-East Asia. Like the eradication campaigns that had gone before, the ethos of the 'let us spray' approach was top-down, with no effort made to improve the lives of the communities afflicted. In a breathtaking development, the supposedly 'global' MEP decided to exclude the entire continent of Africa from its efforts, as it was deemed too difficult to attempt eradication there.

In its first few years the programme seemed to be an amazing success. Eleven countries eliminated malaria by 1960 and twelve saw their malaria rate fall to negligible levels. India, so long plagued by the disease, reduced its caseload from 75 million to fewer than 100,000. Across the world life expectancy and food production rose, just as economists had predicted. Universities closed down their malaria departments and Paul Russell was fêted as 'the man who ended malaria'. The scale of the campaign was enormous. In 1961, more than 66 countries were involved in eradication, harnessing the effort of 200,000 sprayers and targeting 600 million people. But by the middle of the decade the malaria eradication programme was in serious trouble on several fronts and the disease was about to stage a comeback.

DDT might have been a 'wonder chemical' that graced the cover of *Newsweek,* potently illustrated as a mushroom cloud. But ordinary people and environmentalists noticed that DDT had some less than wonderful consequences. It killed not only insects and mosquitoes but also bedbugs, chickens and, in one famous case, the cats who ate the cockroaches in a Borneo village, leading to the much-ridiculed 'operation cat drop', in which British RAF planes airdropped twenty healthy replacement cats, in specially constructed carriers, into the middle of the jungle. In the USA, Michigan State University researchers discovered that all the robins in a 185-acre plot died after eating DDT-laced earthworms. As wariness grew, people began to oppose the constant spraying programme, closing their doors and windows when the sprayers appeared or pretending to be out to avoid taking part.

Although US Army reports had noted the potent effects of the chemical on small animals, the general population only began to take notice after the publication in 1962 of Rachel Carson's seminal book, *Silent Spring*, which drew attention to the overuse of pesticides. Carson did not claim that pesticides should never be used, only that, 'We have poisonous and potentially potent chemicals indiscriminately in the hands of persons largely or wholly ignorant of their potentials for harm.'[10]

This was the beginning of the end for the malaria campaign. In 1963 the USA decided to stop funding it completely, USAID withdrew from the programme and UNICEF halved its malaria staff. Six years later, in 1969, the World Health Assembly directed the WHO to abandon the eradication plan, claiming it could not be successful without new commitments from national governments, which appeared to be nowhere in evidence. The campaign had cost US$1.4 billion.

DDT was banned in the US in 1972 but an over-reliance on spraying as the sole means of combatting malaria meant that DDT-resistant mosquitoes had appeared all over the world by then. Mass medication using a single drug, chloroquine, meant that drug-resistant parasites were simultaneously on the rise, leaving the world with neither an effective preventative spraying nor drug treatment. In Brazil and in other parts of the world, chloroquine had been treated as a wonder drug, even laced into table salt, leaving millions of people vulnerable when drug resistance struck.

As with the other eradication campaigns, the failures were not only chemical but also human. 'In India fewer than one in nine spray teams adequately sprayed their assigned areas', wrote Sonia Shah in *The Fever*. 'In a village of sixty-three houses ten doors would be locked, thirty-five residents would refuse access and one house would be forgotten.'[11] Sprayers became disheartened, surveillance teams skipped hard-to-reach villages and blood samples languished in laboratories for months. A campaign envisaged as a top-down spraying effort faltered as inadequate health systems and poorly trained workers proved inadequate to bring about the rigorous revolution required to eradicate the parasite.

Malaria had fallen from 350 million cases around the world to only 100 million, although that figure excluded the large swathes of Africa that the health community had decided not to treat. Malaria cases had been brought close to zero in India, Pakistan and dozens of other countries. As funding dried up, however, government attention strayed to other matters, the WHO shifted its efforts back to malaria 'control' and the campaign collapsed. Malaria surged back, leaving people in malaria-afflicted areas of the world worse off than before. The morale of those who had supported eradication plunged. Paul Russell was a broken man. Fred Soper neatly expunged almost any mention of the malaria campaign from his memoirs. Without a magic-bullet chemical or drug, those who wanted to tackle the disease in the coming decades would have to find different and more complex, methods. With some degree of success, they did.

A new beginning

'After the end of the global malaria eradication campaign, during the next couple of decades, malaria generally worsened around the world,' said Sir Richard Feachem, the founding director of the Global Fund to Fight AIDS, Tuberculosis and Malaria, who now runs the Global Health Group at the University of California in San Francisco:

> Then in the early part of this century things began to change rapidly. The year 2002 is the most significant date because it's the date for the creation of the Global Fund to Fight AIDS, TB and Malaria, and between 2002 and 2016 we've seen remarkable reductions in malaria all around the world.

Feachem says the turnaround can be attributed to three things: additional money, new and better technologies and renewed ambition:

> Investment jumped and continued to rise from 2002 onwards and that was entirely due to the Global Fund and then followed up by the US President's Malaria Initiative, the PMI [launched by President George W. Bush]. And those two big programmes between them turned investment in malaria from a few tens of millions of dollars per year into billions of dollars per year and that huge increase in investment undoubtedly was one of the key elements.[12]

The new and better technologies Feachem refers to include long-lasting insecticide-treated bed nets, rapid diagnostic tests and Artemisinin-based drug treatment.

Distribution of long-lasting impregnated bed nets treated with insecticide began in 2000 and has proved one of the most effective measures against malaria. The share of the populations at risk of malaria who sleep under a bed net rose from three per cent in 2004 to forty-four per cent in 2013. A recent study in *Nature* by Bhatt, Weiss and Cameron[13] estimated that the nets are responsible for sixty-eight per cent of aversions of clinical cases of disease between 2000 and 2015. After much debate, in 2014 the WHO decided to recommend free universal coverage for bed nets and studies show that extending the life span of bed nets from three years to five could save up to US$3.8 billion.

The development of rapid diagnostic testing has also transformed the world of rural malaria treatment. An article in *The Economist* in November 2015 described the testing and surveillance regime in Swaziland, a country experts predict will soon be free of malaria:

> In a dusty yard in Magagasi, a small village in eastern Swaziland, a man in surgical gloves draws Gugu Dlamini's blood for the third time this year. The health worker lays a drop of it on a small plastic tray and adds a clear solution. The ritual is familiar. Every time a malaria case is reported in the country, surveillance officers sweep in and test everyone living within five hundred metres of the sick person. In a few minutes a single line appears in the tray's indicator window: Ms Dlamini does not have malaria.[14]

Rapid diagnostic tests cannot be used alone. As *The Economist* noted:

> The field tests that detect the parasite in places like Magagasi are not sensitive enough to pick up low-grade infections. Laboratory equipment that can detect the parasites is not available in every country. In Swaziland surveillance officers collect samples for laboratory analysis alongside rapid testing and track them using bar codes and GPS coordinates to help them return to the right house.[15]

Rapid diagnostic testing marks a crucial advance, as people who are treated for malaria are less likely to transmit the parasite and correct diagnosis prevents the overuse of anti-malaria drugs that leads to drug resistance. Bhatt's paper in *Nature* attributed a further ten per cent of the remaining fall in cases to the existing house-interior spraying projects and the final twenty-two per cent to new combination drug treatments involving ACT, which are estimated to reduce malaria deaths in children by ninety-six per cent.

'In the early years of this century Artemisinin Combination Therapy became available and it was rolled out very rapidly to be the first line treatment in almost every country,' Richard Feachem said.[16] The story of artemisinin is truly a story of the geopolitics of the twentieth century, emerging from the depths of the Vietnam War in the mid-1960s and the rise of chloroquine-resistant parasites.

The medicinal properties of *Artemisia annua* (the sweet worm-wood tree) have been documented in ancient China since 168 BCE.

The tree contains a fragile compound called artemisinin, which kills malaria parasites; in 340 CE the physician Ge Hong gave specific written instructions on how to use the plant, soaking it whole then wringing it out and using the juice to provide relief from fever. This knowledge remained untapped until Chairman Mao instructed Chinese doctors and scientists to find an alternative to quinine, which was in short supply, as a treatment for soldiers in Vietnam.

The scientists for the top-secret Project 523 gathered in May 1967 at a Beijing restaurant to search through traditional texts for clues to a new cure for malaria and Ge Hong's writings came to light. However, the scientists did not follow the instructions accurately, killing the artemisinin by drying or heating the leaves. The process was discarded as useless until it was rediscovered, more successfully, in 1972. Discovering that artemisinin killed malaria parasites more quickly and effectively than either quinine or chloroquine, Chinese scientists quickly put the drug into use, cutting the death toll for Vietnamese and Chinese troops by a third in the final stages of the war.

For obvious reasons the 'scientific fairy tale'[17] of artemisinin remained a closely guarded Chinese military secret. The first details did not appear in English until 1979. Although the drug slashed malaria cases in China from 2 million to 90,000, it remained largely ignored in the rest of the world, left out of academic papers and looked down on by Western scientists who scoffed at Chinese methods. Not until 1994 did a major pharmaceutical company, Novartis, begin producing the first artemisinin-based drug. It was a further five years before a combined artemisinin drug (ACT), which proved more effective in combatting drug resistant parasites, was launched. Even then it was a highly expensive treatment, costing US$44 for one course, while the chloroquine alternative cost only twenty-five cents. In 2015, Youyou Tu, who led the arte-misinin project in the 1960s and 1970s, became the first Chinese scientist to win the Nobel Prize for Physiology or Medicine. Although her recognition was long overdue, the role of the drug treatment she discovered may have been eclipsed before its potential was fully realized.

'Those were three very significant technology improvements, which undoubtedly were a key driver of progress,' said Sir Richard Feachem, 'but the third, I think, is a much more subtle one and it's to do with ambition and commitment.' When Feachem began leading the Global Fund in 2002 he noticed a 'kind of fatalism about malaria', a sense that the disease had always been with humankind and always would be. Yet, in the course of five years, that attitude changed quite dramatically:

> It really struck me how countries were doing better and noticing that they were doing better and getting more ambitious. So some countries with high burdens of malaria began to see a day when they would have very low burdens of malaria and some countries that already had quite a low burden began to talk about elimination. There are two 'e' words, elimination and eradication and those two 'e' words could not be used in polite company in the 1980s or 90s. The 'e' words were not acceptable. But that began to change and countries began to think that elimination might be possible and that process continued and gained momentum.

The Malaria Elimination Initiative that Feachem began work on at UCSF in 2007 was part of the early wave of the commitment to elimination, he believed, although the tipping point in favour of eradication was not reached until 2015. Feachem points to the September 2015 report *From Aspiration to Action*, produced by Bill Gates and Ray Chambers, the UN secretary general's special envoy for malaria, as a clear call for eradication. 'The second "e" word is on the table, and an increasing proportion of people working in malaria think that it is possible,' Feachem said. 'This region by region elimination, leading to global eradication, is now mainstream thinking, whereas as recently as 2007 it was very radical thinking.'[18] The report estimated that eradication could save eleven million lives and unlock trillions of dollars in economic benefits. To do so, however, will require a new range of technological and logistical transformations.

In Africa, mosquitoes have already become resistant to pyrethroids, the chemical used to treat bed nets and used in two-thirds of house spraying. Studies have also shown that take-up of bed nets is less than is sometimes quoted, with people preferring to sleep outside or using the nets incorrectly. Another key weapon in the arsenal is also under threat: artemisinin. The WHO did not recommend ACT as the first drug to be given for malaria until 2001. Even when Novartis dropped the price to (a still relatively expensive) two dollars, USAID, the CDC and United Nations Children's Fund resisted the use of ACT in favour of less effective chloroquine treatments until 2004. In the intervening years a huge underground market in artemisinin sprang up, drowning Africa in less effective copycat drugs that used artemisinin as a single agent, making it easier for the parasites to develop drug resistance. Artemisinin-resistant parasites are currently present in several South-East Asian countries, including Cambodia, Laos, Myanmar, Thailand and Vietnam. While many people with this strain have recovered after treatment with other combination drugs, some parasites have remained resistant to all treatment. The era of ACT may be over before it has even fully begun.

Wiping out a disease as complex and adaptable as malaria requires funding, innovation and the kind of logistical rigour that the polio campaign demonstrated in its final years. In the ten years after it was founded in 2002 the Global Fund spent $8 billion on malaria, pooling resources from government, charities and the private sector. While in the past countries in the developing world often shouldered most of the burden for disease eradication, the money from charities and wealthy governments now accounts for eighty-two per cent of the US$2.7 billion spent on control and elimination in 2013. The President's Malaria Initiative, launched by George W. Bush, has an annual budget of $674 million. Economists estimate, however, that is still a fraction of the total cost of eradication, which could total $120 billion by 2040, peaking at $6 billion a year. That money will not only be needed for a multi-pronged campaign but also to fund research in new drug developments, possible vaccines and better disease surveillance.

The elusive vaccine

After three decades of research it appears that a vaccine – the much-discussed RTS,S developed by Glaxo Smith Kline and the Malaria Vaccine Initiative – may be close. Unlike the polio vaccine, which is ninety-six per cent effective, RTS,S cuts the number of malaria cases by only thirty-six per cent over four years for infants and very young children. Experts estimate that it is about half as effective as an ideal vaccine should be to support eradication.

'To have a high efficacy, easy-to-administer vaccine at a reasonable cost would be a huge asset but we still don't have it,' Richard Feachem said:

It wouldn't stop us doing all the other things
we do but it would be a very powerful additional tool in the toolbox. The key question is,
do we need one to finish the job? I firmly
believe we can get there without it but there's
no shortage of people who don't agree with
that.

Having worked across malaria and HIV, Feachem draws a parallel
between the two diseases. A vaccine for HIV has proved incredibly
elusive, with no effective candidate appearing, even after decades
of research; in comparison, drugs to treat the disease get better
and better. Feachem said his hunch is that the end of HIV will be a
solution based on drug treatment, not a preventative vaccine:
'I think broadly speaking the same is true of malaria. The end of
malaria will be a drug-based solution, not a vaccine-based solution.'[19] Mass campaigns, using anti-malarial drugs, proved effective
in villages and regions in Asia, including a 1991 campaign in
Vanuatu, where nine rounds eliminated the disease before it made
a comeback a decade later. More recently, between 2011 and 2012,
cases fell from 209 to 46 on the Thai-Myanmar border.

With the influx of new funding, the focus of research on
malaria drug treatments is to find a replacement for artemisinin.
Although several drugs appear promising, none will be available
for years. According to Richard Feachem:

> The dream is a single-encounter radical cure
> for all malaria species. So the goal of the drug
> research is to be able to have a single encoun-
> ter with an infected individual and in that
> single encounter administer a pill or pills

which would cure any of the four or any of the five plasmodium species that that person might happen to be infected with. Such a drug or drugs – it may well be a combination of drugs – would be a massive breakthrough and research is very focused on getting as close to that as quickly as possible.[20]

David Schellenberg of the London School of Hygiene and Tropical Medicine and director of the ACT Consortium that looks at malaria drug delivery, explained what single encounter means: 'As we move towards eliminating malaria it would be really useful to have drugs which not only clear active infections in people but also provide a period of protection from the disease afterwards.'[21] This is known as SERCAP (single encounter radical cure and prophylaxis). Such drugs would be particularly effective when disease surveillance teams detect a large number of cases in a particular location and decide to administer drugs en masse. A SERCAP would, Schellenberg believes, offer more incentives to pharmaceutical companies, because programmes would still need to buy and give large numbers of anti-malarial drugs, rather than a single vaccine, which produces fairly low returns for drug companies.

A second crucial area of drug development is targeting the *Plasmodium vivax* strain of malaria. The impact of this strain, in terms of causing severe illness and death has, Schellenberg says, been underestimated but the existing drug used to treat *vivax* can cause fatal haemorrhages. *P. vivax* is prevalent in South-East Asia, notably in a border area between Thailand and Cambodia that has proven to be a hotbed for drug resistance. Malaria elimination in this region will be crucial to any malaria eradication campaign.

One of the problems with *vivax* is that it remains dormant in the liver, post-infection. Schellenberg explained:

> At the moment we don't have any tests for people who have got that dormant liver infection and it's not really clear how one could go about developing a test for the dormant liver stage. That's important because at the moment there's only one drug which is recognized to kill the dormant liver form and that's the drug primaquine. The two unfortunate things about that drug are that it needs to be taken for fourteen days and that it can cause the rupture of the red blood cells in people who have an inherited disorder of their red blood cells, G6 PD deficiency, which is relatively common in places where there's a lot of malaria. The problem is that if you then take primaquine you're at risk of haemolysis – rupturing of the red blood cells – and that can be life-threatening.[22]

In addition to drug developments, Richard Feachem also believes that the newly vibrant research sector will lead to more advanced rapid diagnostic tests and more effective vector – mosquito – control. 'I think quite soon we'll have a supersensitive test which will have a big impact in the field,' Feachem explained:

> The research is focusing on much, much more sensitive rapid diagnostic tests, which can pick up low levels of parasites in the blood

and can recognize all parasite species. We're focusing on the sensitivity, not the specificity. We don't want false negatives. If you are infected, even at a very low level, we want to know that. If we got a false positive, we wouldn't mind too much about that. We would treat you with a very safe drug and there would be no bad consequences from that.[23]

The first public health insecticide

While vector control has been a 'dormant area' for years, new partnerships between private industry and research laboratories have led to the discovery of new classes of safe and effective insecticides. Janet Hemingway, professor of Insect Molecular Biology and director of the Liverpool School of Tropical Medicine, said:

We've got two main interventions, which are the long-lasting insecticide treated nets and the indoor spray, that have been around for the last twenty to thirty years but innovation in that area had largely stopped until about ten years ago and it had stopped in terms of us getting no new public health insecticides onto the market.[24]

Insecticides used in public health were traditionally developed from agricultural insecticides but changes in agricultural practice, away from contact insecticides sprayed on crops, means that no agri-chemicals developed in the last thirty years are suitable for public health purposes. Janet Hemingway noted:

> Although vector control accounts for some-
> where between sixty and eighty per cent of
> the reduction in malaria transmission, unless
> we were going to develop new insecticides
> and create a brand new pipeline for public
> health insecticides, then we were going to be
> in trouble very soon . . . We would get resis-
> tance to the compounds that we already have
> and have nothing to replace them with.

The Innovative Vector Control Consortium (IVCC) was set up eleven years ago to develop public health insecticides, creating new formulations of existing chemistry for indoor residual spray-ing and new routes to market for new public health insecticides, three of which have been selected to go through to full devel-opment. These are, Hemingway stressed 'a brand new class of insecticide which are not affected by the resistance that we have got to the current insecticides that are out there'. These new prod-ucts will take from seven to ten years to reach the market. New regulatory pathways must be set up 'because nobody has ever developed an insecticide just for public health before'. Introducing new bed nets also raises tricky issues, Janet Hemingway said: 'In terms of the nets, the push there has been to try and drive the price down to try and get universal coverage. But at the moment those are completely reliant on one class of insecticides, which is just starting to fail in Africa.'

A 2013 report by NPR revealed what Janet Hemingway had long predicted: families in the village of Chikwawa, in southern Malawi, faced the devastating consequences of relying on bed nets that used only one type of insecticide. In Chikwawa, the lake that lies just beyond the village is a breeding ground for mosquitoes but until recently, local people put their faith in their bed nets. Now, small children, such as the one-year-old daughter of Alice Sekani, were lying sick and immobile in hospital, suffering from a potentially fatal form of cerebral malaria. Yet Alice Sekani said her daughter slept under a bed net every night.

Themba Mzilahowa, from Malawi's Malaria Alert, tests resistance by collecting mosquitoes in a tube, exposing them to insecticides and seeing if they survive. Until 2010, every mosquito he tested was killed by pyrethroids. But only three years later, the results were very different: 'We're seeing failure of the chemical to knock down the mosquitoes,' he said. 'We have to get new insecticides on to those nets and make sure that those get into the market,' Janet Hemingway urged, but there is a conflict between keeping the goal of universal coverage and introducing new more expensive nets. 'We face a conundrum because the new insecticides are going to be more expensive but we still have to supply the bed nets at a very low price.'[25]

Work is under way on many other methods of vector control, including repellents, sugar-baited traps, eave tubes with insecticide netting and impregnated wall linings, but cost and scale remain big hurdles.

Malaria eradication 2.0

Drugs, vaccines and insecticides are the existing tools for eradicators, Richard Feachem pointed out, but new tools, such as better software and surveillance mapping will be just as crucial in future. He remains cautious about genetically modified mosquitoes: 'How do you actually replace the wild population by your laboratory designer mosquitoes? I don't think people talk enough about that challenge,' Feachem said. Producing a modified mosquito in the laboratory is a first step but:

Then you've got to find a way to actually release huge numbers of your 'designer mosquitoes' and have them be more fit, in the Darwinian sense, than the wild population. Otherwise they'll fly around for a while and die out and the wild population will occupy the space once again and that's a problem that has not yet been solved.[26]

Nor should the political difficulties of introducing wild genetically modified mosquitoes be underestimated, according to Janet Hemingway:

It will be a big political decision as to whether or not they really get deployed because mosquitoes don't carry passports. So if you decide you're going to release these into Africa then who needs to agree and how do you get them to agree? If the technology works, releasing them in one country means that they are going to spread to other countries. I wouldn't underestimate the work that is going to need to be done to start and use these as large scale releases in big areas of the world.[27]

Nonetheless, Richard Feachem pointed out: 'We have today's tools today, those are the ones we use in one hundred countries around the world and they're very effective and they're driving down malaria and they are, generally speaking, working. But science marches on and over the next few years the tools will improve . . . and we will use them.'[28]

Much of the work of a good surveillance campaign is done by the drudgery of foot soldiers but Swaziland is bringing together traditional methods and Google-powered data analysis. When a person with malaria appears at a local clinic a nurse rings an emergency number that automatically sends a text message to malaria-control workers. Their work tracking down the precise location of the transmission is supplemented by new software developed by Hugh Sturrock, working under Richard Feachem at UCSF, which uses Google Earth data to analyse weather and land-scapes to pin down which villages are most at risk. Hugh Sturrock explained the rationale behind the project:

> Malaria is a very patchy disease, it's not every-where. It's clustered in certain countries. Even within countries it's in certain places and not in other places. And even in villages some households experience malaria many times in the year while some never see it. We can predict where malaria might occur and so we can target our resources to places where we think malaria is, rather than wasting a lot of money distributing things like bed nets and insecticide in places that really are at very low risk.[29]

As a spatial epidemiologist, Sturrock is used to working with maps but points out that maps can only present a single, static view:

> You find that you get a single map that shows you the picture of malaria at that point in time but malaria can vary over time and it's

very seasonal. Hotspots can move around. So what we really need is a mechanism or a platform that is able to produce rapidly updatable maps, almost like weather maps of malaria, so that we can get ahead of the disease and we can target it and hit in the places that are being most affected.'

With funding from Google, Sturrock discovered that Google Earth imaging and the Earth Engine enabled him to produce the kind of 'malaria weather maps' he was looking for:

A lot of these datasets are open and freely available online but they're scattered all over the web and through different agencies and in different formats; they're difficult to use. So what the Google Earth Engine has done is compile all of these data layers and put them together in a spatial data warehouse so you can visualize and manipulate the data. You can then create summaries and use Google's computers to do computations.

In seconds, Google's Earth Engine can provide an answer that, Sturrock estimates, would previously have taken weeks to download and compile. The team has been piloting the new tool in Swaziland, taking the GPS location the malaria team provides for each case and plugging it into the new platform, Sturrock explained:

It looks at all of the cases and where they've occurred relative to the levels of rainfall at that point in time or the density of vegetation or the temperature or the elevation or the slope or the distance to water bodies and other variables that we know are related to malaria because they're related to mosquito patterns of distributions.

Algorithms then work out where conditions are most conducive to malaria transmission, and that information can be turned into a statistical modal to make predictions elsewhere: 'We take case data and convert it into what we call a risk map – like a weather map – for where we think malaria is occurring.' Next, the team plans to prototype their work in Zimbabwe, although much depends on having a repository of good existing mapping and data, something that still does not exist in all parts of the world.

Sturrock has found that the shift from malaria control to malaria eradication means that his work is 'brushing up against' the work of the polio campaign:

Polio teams are very good at the logistics of getting to houses and mapping out houses where they should be targeting. Until now that hasn't really been an issue for malaria because it's generally been thought that universal coverage is the way forward. So we don't need to know where individual houses are, we just need to know that in this district ten thousand people require bed nets and we go and give them out. But for malaria elimination you really have to shift to thinking about individual houses. Understanding exactly where people are living becomes really important.

The eradication team needs basic information from organizations such as Médecins Sans Frontières and British Red Cross 'Missing Maps' projects, which organize teams to go street to street and collect data in places where no maps exist, including the DRC and South Sudan. Those data are relayed to volunteers in cities around the world who meet once a month to share pizza and build maps online. 'Volunteers can log in, look at images and draw around building footprints for us and then a second wave of volunteers comes along and verifies and validates those data layers,' Sturrock explained.[30]

Ivan Gayton, the Missing Maps project coordinator, said:

> We get our aerial imagery and then bunch of people in London and Vancouver and Jakarta and Senegal have a mapping party and they trace it. Then we have a map without any names on it, so we go to the field, we speak with the local people and we actually collect names of streets and neighbourhoods and villages and we put it together into a final map.[31]

Gayton is using data from Lubumbashi, in the southern part of the DRC, as a test-case, creating a dashboard for real-time data and monitoring of cholera; something that could be applied to other diseases.

'Rolling it back'

Public health workers like to produce a chart which shows malaria rolling away, across the globe, from country after country. Seamlessly, the coloured disease-affected areas slip away, leaving larger and larger patches of a white, disease-free landscape. At the turn of the twentieth century malaria was present in two hundred countries; today it remains endemic in one hundred. 'Of those one hundred, over a third are in elimination mode. In other words, they're going from low to zero and they're making rapid progress,' Richard Feachem said:

> The other countries are a little behind that. They're going maybe from high to medium or from medium to low but of course they will eventually quite soon get to low and then switch their programmes from control programmes to elimination programmes and have the target of zero. So it's a very rapid, exciting and encouraging progress, which we're right in the thick of now.[32]

Sceptics point out that the roll-back map does not show that many of the countries – such as the UK – rid themselves of malaria through population movements and improvements in standards of living and a host of other factors unrelated to mosquito control campaigns. Nor does the map convey the complexity, or the tenacity, of malaria in those spots in the world where it clings.

As the failed malaria eradication campaign of the 1960s and other failed single-disease campaigns demonstrate, complexity and tenacity are often the biggest enemies of eradication efforts. Especially efforts that must be fuelled and driven by a community of different groups, drawn from government, private groups and charities, who may lose interest, balk at rising costs and lack the will to maintain funding for a long and arduous campaign. This time, however, Richard Feachem believes, it is different: 'There's been a dramatic movement of the consensus, if you like. The consensus is now certainly pro-elimination and largely pro-eradication as well, whereas those views were the fringe, the extreme fringe in 2007.'

Feachem cites the transformation in China, a country he frequently visited during his work at the Global Fund. 'In five years, China moved from a not very ambitious controlled programme to a determination to eliminate,' Feachem said. On every visit, he noticed Chinese ambitions rising until, in 2010, the country declared its goal was elimination by 2020. Malaria cases were reduced from 26,000 in 2008 to 2,716 in 2012, of which only 243 were due to local transmission. Feachem commented:

> China only had about fifty cases of indigenous locally transmitted malaria last year and they were all in a very few districts, either in the province of Yunnan on the border with Myanmar or in Tibet, on the border with India. That's all that's left of locally transmitted malaria in China, which in recent historical times used to cause many millions of cases a year.

New money, new tools and new ambition are the three factors responsible for the dramatic driving down of malaria. Now, Feachem conceded, for eradication to be achieved, the loose partnerships of organizations and research interests must come together in a tighter coalition. As yet the malaria eradication movement still cannot be called a campaign. 'Is it a campaign or is it a movement with many players?' Feachem queried:

> Campaign sounds very unified, organized and coherent and this is not unified or coherent. This is not a single well-oiled machine moving forward in lockstep at all. I would say it's more of a movement at the moment, rather than a campaign and there are many key actors playing different roles.

Key players include the Gates Foundation, Ray Chambers and the team at the UN, the WHO, the Global Fund and national governments, including the US (through its contributions to the Global Fund and the President's Malaria Initiative), the Australian government, particularly in the Pacific, and most recently and to everyone's surprise, the UK, which recently announced £3 billion of funding over five years.

Feachem called the announcement, made at the Liverpool School of Tropical Medicine by the former chancellor of the exchequer George Osborne, at an event co-hosted with Bill Gates, 'a dramatic shift in policy'. The Department for International Development and the UK government have long been involved in malaria research but 'there has not been a UK government commitment to malaria eradication, to ending malaria, using words of that kind, until Liverpool'.

Driving eradication forward means moving into a better-organized and better coordinated campaign to create the kind of global structure that brought together the polio eradication effort in its final stages. Just as in polio, that will ruffle some feathers.

Creating a global structure involves rethinking the roles of the major players, including governments, international bodies, the Gates Foundation and other charities, and their relationships with the governments of countries where malaria is still endemic. 'It's an international architecture that is not just among the Northern players and about channelling money from the North to the South,' Feachem said. 'It has to be an international architecture that involves the endemic countries as well and organizational structures that they already have.'[33] Once the new structure is agreed and the new global 'malaria eradication campaign' emerges, it can go forth to the World Health Assembly for endorsement by 2020, if not sooner.

That route has been taken many times before, for many different diseases, but who would predict failure when the eradicators have shown themselves to be just as tenacious, determined and adaptable as the mosquito? 'No animal on earth has touched so directly and profoundly the lives of so many human beings', wrote Andrew Spielman in *Mosquito, The Story of Man's Deadliest Foe*. 'For all of history and all over the globe she has been a nuisance, a pain and an angel of death. Mosquitoes have felled great leaders, decimated armies and decided the fate of nations. All of this and she is roughly the size and weight of grape seed.'[34]

DO NOT VACCINATE!!

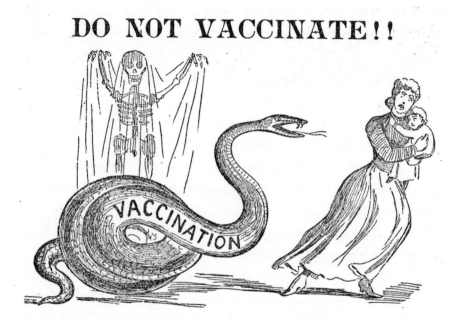

The anti-vaccination movement stretches back hundreds of years. As this advertisement shows, those who feared the 'poisoned lancet' were as vocal in the nineteenth century as they are today.

7

LIBERTY OR DEATH – THE ANTI-VACCINATION MOVEMENT

A particularly filthy bit of witchcraft. . .
Anti-vaccination campaigner, George Bernard Shaw[1]

They called it the 'Mickey Mouse measles' but the so-called 'happiest place on earth' became anything but that for the people who caught measles there. In the week leading up to Christmas 2014 thousands of families paraded up and down Disneyland's Main Street in Anaheim, California. One of those people, probably someone who had recently travelled abroad, carried a strain of the measles virus (B3) that had been prevalent in an outbreak that struck the Philippines earlier that year. By 7 January 2015 officials were warning that the Disneyland visitor was linked to seven confirmed cases of measles in California. Two people in Utah, and six of the people infected from California, were not vaccinated against the disease. By the end of January 50 people in California had been infected by the Disneyland outbreak, which eventually spread to 140 people across the USA, Canada and Mexico.

War, politics, technology and logistics have conspired to make disease eradication difficult, if not impossible but no factor has played a bigger part than the refusal of people to be vaccinated. Bad-tempered – and in the past violent – rows about vaccination are usually presented as a battle between science and the unenlightened masses. In reality, they are power struggles, in which science plays a very small role; the same power struggles that have consumed the world for the last two hundred years: struggles between races, between classes, and between men and women. Anti-vaccinators have compared their cause to the abolition of slavery, or the fight for Home Rule in Ireland, while women have picked up the refrain, asking in essence the same questions. What rights do we have over our bodies? As mothers, what right do we have over our children's bodies?

The history of disease eradication in the twentieth century was driven by imperialist and colonialist urges; of white men chasing local women around their huts and jabbing them against their will but supposedly for their good; of spraying their houses until they dripped with DDT; of returning to a family again and again until, worn down, a mother agreed to give her child OPV. In other places and times, vaccination was driven less by imperialism than by outright racism or the heavy hand of the state clamping down on the lower classes. In her book *On Immunity* Eula Biss pointed out that in the 1898 smallpox outbreak in the US some believed white people could not catch the disease, calling it the 'nigger itch'. In Middlesboro, Kentucky, black people were vaccinated at gunpoint.[2]

The Disneyland measles outbreak reignited a highly contentious, and highly political, debate about disease, vaccines and the power of the state versus individual rights. More than 640 people were infected with measles in the US in 2014, the highest number for twenty years, and more than in the previous four years combined. Worse came when, in the summer of 2015, a woman in Clallam County, Washington State contracted measles in an out-break linked to a local health centre, later dying from pneumonia brought on by the disease. She was the first person in America to die from measles for twelve years.

'We hope to see the day when no one dies from measles,' Anne Schuchat, director of the National Center for Immunization and Respiratory Diseases at the CDC, told the Associated Press. 'We rely on people who can be vaccinated to protect those for whom vaccine protection isn't available.' The unprotected include people such as those with leukaemia or those receiving cancer treatments, who cannot be vaccinated against measles or may not have a good immune response even if they have. Or people such as the unnamed woman in Clallam County, who had a depressed immune system.

This is the concept of 'herd immunity', the basis of all vaccin-ation campaigns. Herd immunity relies on the majority of people being vaccinated and thus protecting the minority. It is the opposite of individualism: you may never get the disease or if you do it might be in a mild form so you survive with no ill effects. But by being vaccinated you boost the immunity of your group and protect the vulnerable people within it, who would otherwise be seriously affected by a disease. In other words, you are possibly helping yourself but importantly, you are performing a public service.

For measles, herd immunity is particularly acute. According to the CDC, measles is the 'most deadly of all childhood rash/fever illnesses'. Those mildly infected might have a fever, rash and cough but those with serious infections face the possibility of deadly com-plications. As the Disneyland outbreak demonstrated, the measles virus is highly contagious and easily transmitted when an infected person coughs or sneezes. Up to ninety per cent of those who are not immune will become infected if someone close to them has measles.

Frustratingly, for doctors and public health workers, although a highly effective measles vaccine has been widely available since the 1960s, up to twenty million people around the world still contract the disease every year. Numbers are rising in the USA and Europe, due to the spread of the disease among groups of unvaccinated people. While some shun the vaccine for religious reasons (a large outbreak occurred in Pennsylvania's Amish community a couple of years ago), a growing number of people have refused the vaccine as a result of Andrew Wakefield's 1998 study published in *The Lancet*[3] that described an unfounded, and now completely debunked, link between the triple measles, mumps and rubella (MMR) vaccine and autism.

That article led to an immediate and dramatic drop in vaccination rates in the UK and Europe. The furore even surrounded the then prime minister's son, Leo Blair, when journalists demanded to know if he would be receiving MMR (he did but the Blairs' decision wasn't made public for years). In 2010, *The Lancet* fully retracted the Wakefield study, describing it as 'utterly false'. Andrew Wakefield was found guilty of serious professional misconduct and struck off by the UK General Medical Council. Nonetheless, the consequences of his study live on. In California, when the Disneyland outbreak reopened a dispute about whether vaccines should remain compulsory for school-age children, fears over MMR fed into a bigger debate about individual rights, pitting parents against politicians.

Although all American states mandate vaccination for schoolchildren, all allow medical exemptions and twenty allow exemptions on grounds of religion or personal belief. While accepting that medical exemptions should remain, the governor of California, Jerry Brown, believed that exemptions on the grounds of personal belief had allowed large groups of parents (usually presented as from the middle class) to opt out of vaccinating their children because of the ongoing controversy over the MMR vaccine. Up to eighty thousand California children had not been vaccinated on the grounds of their parents' personal belief; vaccination rates in some suburban areas were as low as fifty per cent. This weakened the state's herd immunity to the disease and contributed to the widespread Disneyland outbreak.

After an emotional and hotly contested campaign, on 30 June 2015 Governor Brown signed one of America's strictest school vaccination laws into law. California joined Mississippi and West Virginia in banning vaccination exemptions based on religious grounds. This ban was widely championed by the medical profession: Dr Paul Offit, chief of the division of infectious diseases at Children's Hospital of Philadelphia, called the ruling 'A great day for California's children' and told residents: 'You're living in a state that just got a little safer.'[4] Although polling showed that sixty-seven per cent of parents agreed with the ruling, opponents such as Rebecca Estepp, who claimed her son had been injured by a vaccine, described the decision as 'heartbreaking' and added, 'It's so coercive. It's so punitive.'[5] Other parents claimed the ruling undermined their children's constitutional right to an education and threatened to leave the state.

Compulsory vaccination had been argued before the Supreme Court in 1922 and 1944 and in both cases the vaccination laws were upheld. But the anti-vaccination movement is far from being a twentieth-century phenomenon; vaccination laws are some of the most contentious in history.

'A poisoned lancet?'

In *Angel of Death*, Gareth Williams described the anti-vaccination movement as a 'two-hundred-year war that shows no sign of ending'.[6] As Jenner knew to his cost, opponents of his great discovery sprang up almost immediately. The first were fellow doctors, jealous colleagues or adherents of variolation or miasma theory, who feared that the new vaccine would put their livelihood at stake. Suspicions and concerns soon spread more generally throughout the population, especially when vaccination was forced upon communities with punitive and heavy-handed zeal.

In England and Wales a series of Vaccination Acts published between 1840 and 1871 led to fierce opposition and dissent. There were outbreaks of rioting in some towns and parents were repeatedly sent to prison for refusing to vaccinate their children. Parents feared, with some reason, that vaccination was a cause of other diseases and that their liberties and rights were being undermined by the Acts' draconian measures. The first Act to Extend the Practice of Vaccination (1840) made variolation illegal and encouraged parents who had not had their children vaccinated against smallpox to do so. When that failed to lead to sufficient levels of vaccination, a revision in 1853 made vaccination compulsory; children had to be taken to a vaccinator within three months of birth and parents were fined twenty shillings if they refused to comply. Three more revisions of the Act followed, in 1858, 1867 and 1871, giving vaccinators the right to collect lymph from any vaccinated child and appointing Town Guardians to enforce the law.

The law was applied harshly and desperate parents sought recourse. A father in Farringdon appeared in breach of the law thirty-two times, even though he said he feared the vaccine because a neighbour's child had been disabled by it. A woman tore up floorboards with her bare hands and drowned herself and her baby in the water storage tank below, rather than go through with the procedure. Some parents did not register the birth of their children and others moved around the country, keeping one step ahead of the Guardians. The Act seemed to weigh most heavily on the poor, who could scarcely afford a fine and had no funds to move to another town. A Manchester doctor, John Scott, wrote: 'There is no getting over the fact that vaccination is hated, amongst the working classes in Lancaster at least.'[7]

Although poor parents often refused vaccination for their children, a more organized opposition grew, under a middle-class leadership of doctors, natural healers, clerics and writers, including the writer and playwright George Bernard Shaw. Pressure groups, such as the Anti-Vaccination League and the Personal Rights Association, published pamphlets, organized meetings and celebrated the cases of parents who had refused the vaccine. In Leicester, a hotbed of anti-vaccination sentiment, twenty thousand people met to publicly burn the Vaccination Acts.

The language employed could be fevered. An engineer and prolific pamphlet-writer, C. Godfrey Gumpel, warned of the fear that drove parents to seek protection from 'the poisoned lancet'. An ardent political opponent of vaccination and Member of Parliament, P. A. Taylor, invoked the image of the 'greasy-heeled horses' and the 'cow's dirty nipples' that were part of the process of making the vaccine, even though that method was no longer used. In a statement to the House of Commons in April 1879, Taylor said he had 'seen dozens and scores of persons who had stated to him that they honestly believed that their children had died from Vaccination,' and that those children had 'died in agony'.

Although much of the methodology and wild claims of the anti-vaccine movement were wrong, there was some truth in their claims. While some opponents could be dismissed as ill-informed, driven by self-interest or just plain cranks, there was a measure of truth in some of the concerns raised by doctors. The failure of vaccination supporters to consider that they might be wrong in any respect set back their cause by decades.

Vaccination was primarily opposed on two medical grounds: first that vaccination itself did not stop people getting smallpox; second that the vaccination infected people with other diseases. Debate raged for years about whether the smallpox vaccine prevented the disease. Two inflammatory studies formed the basis of the argument. One, produced by Dr Keller, chief doctor to Austrian Railways, after a smallpox outbreak in 1872–73 showed that vaccinated people had a higher mortality rate from smallpox than those who had not been vaccinated. This study created great controversy, but it was subsequently found that Dr Keller had made up his findings and that the reverse was actually true. Although the study was eventually laid to rest, it proved the saying that a lie often travels halfway round the world before the truth can put its boots on.

More worrying was a study produced by a former believer in Jenner's work, Professor Vogt, of the Hygiene Institute in Bern, who studied forty thousand smallpox deaths. He claimed to be shocked to discover that most of the cases were in people who were vaccinated. However, experience showed that some of those people either lied about their vaccination or believed they were vaccinated but the vaccines had not worked. Perhaps the figures would have made more sense if Vogt had said that the deaths occurred among people who *claimed* they were vaccinated.

Vogt's work, and other data gathered from smallpox outbreaks, clearly pointed to something that Jenner's fans refused to countenance: his vaccine wore off and needed to be repeated. By 1870 the rate of babies vaccinated against smallpox in England and Wales had risen to between ninety and ninety-seven per cent but a significant number of older people in the community had never been vaccinated at all, while others had been vaccinated twenty years earlier and the vaccine was wearing off. Contrary to what Jenner had insisted, a single vaccine did not last for life.

The second damaging allegation levelled against the smallpox vaccine was that it was both harmful in itself, and caused other diseases. In 1880, the International Anti-Vaccination League wrote: 'They may be born of healthy parents, yet they must be exposed to suffering and possible death, through this system of universal State blood poisoning.'[8] Opponents claimed vaccination caused diseases as varied as tuberculosis, cancers, blood poisoning, diphtheria, syphilis and erysipelas, an infection of the skin and soft tissues. The claim that vaccination caused syphilis circulated for decades; it was not finally dispelled until arm to arm transmission was banned and use of any human substance, such as lymphatic fluid, was banned in vaccine production.

One of the most implacable opponents of vaccination was Lora C. Little from New York. Her seven-year-old son Kenneth died in April 1896, directly, his mother believed, as a result of receiving a smallpox vaccination: 'It was as though he had died the day his arm was punctured.'[9] In fact, Kenneth did not die that day but several months later, after first contracting measles and then diphtheria. Nonetheless, Lora Little continued to believe that the vaccination was the root cause. Understandably devastated by the death of her only child, Lora Little began a lifelong crusade against vaccination that took her to Europe and back and led her to found *The Liberator*, a monthly journal dedicated to the cause. Her self-published book, *Crimes of the Coxpox Ring*, was quoted by anti-vaccinationists until the late 1950s. In *Crimes*, Lora Little listed three hundred examples of children she claimed had been maimed and killed by 'vampires fattening themselves on the life-blood of children'; vaccinators, who went on to cover up their crimes.[10] Little described one girl who fought off the vaccinators, and a policeman flourishing his club, as they tried to inoculate her against her will. The child later succumbed to blood poisoning.

In truth, vaccination was a dirty process; a truly 'poisoned lancet'. By the mid-nineteenth century vaccine production was a lucrative business, largely carried out in farmyards, with little purification of the material between its extraction from the cow's udder and its eventual injection into a human arm. The same dirty needle, superficially wiped clean of blood, was frequently used to inflict that injection on thousands of people. Vaccination spread bacteria, including the bacterium that caused erysipelas. Improved reporting in the later part of the nineteenth century showed that half of all deaths attributed to vaccination were actually due to erysipelas. An enforced vaccination of Union prisoners held at Fort Sumter during the American Civil War led to two hundred deaths and hundreds of gangrenous infections.

In the face of evidence that vaccination did indeed kill a very small proportion of children, Queen Victoria set up a Royal Commission in May 1889 to finally settle the question of the safety of the procedure. The commission met 136 times over seven years and eventually produced a majority opinion – with two dissensions – that the decline of smallpox mortality was due to the 'protective influence of vaccination'. The commission also suggested revising the Vaccination Act, however, in the face of much evidence that it was not working; at least one-fifth of local authorities did not enforce the Act and organized hostility took hold in towns such as Leicester and Gloucester. Leicester proved to be an interesting exception to the rule, as its clever system of isolating cases, vaccinating those close to an infected person and imposing good hygiene, meant that not only did it hold off a smallpox outbreak but became a model for the later stages of the smallpox eradication effort in the 1970s. Gloucester, with none of those measures, fell victim to a major epidemic in 1895 that claimed four hundred lives, two-thirds of them small children.

In the face of widespread popular opposition, the commission suggested that those who were fined for refusing the vaccine should no longer have a criminal record and that those who could plead their case of 'conscientious objection' could be excused from vaccination. As the experience in California demonstrated a century later, an ever-broadening definition of 'conscientious objection' meant that the Vaccination Act itself became worthless and was eventually repealed in 1909.

The contested origins of HIV

As the twentieth century progressed vaccination continued to attract attacks from a wide range of opponents, including a strain of Protestant religion that believed the fate of children should be left in God's hands, homeopaths and natural healers who wanted to promote their own cures, and conspiracy theorists who believed that vaccines were either part of a sinister plot or the origin of dreadful mistakes and subsequent cover-ups. The most notorious of these theories was the suggestion that Hilary Koprowski's polio vaccine had caused AIDS.

The polio vaccine was already the subject of controversy, aided by the catastrophe at the Cutter laboratories that resulted in the deaths of ten people, and the incontrovertible fact that as wild polio was eliminated from countries such as the US and UK the only new cases of paralytic polio were caused by taking the live, attenuated oral polio vaccine. In the 1990s, however, the hotly contested issue of vaccine safety was shaken by an even more serious allegation: an article in *Rolling Stone* by a journalist, Tom Curtis, that claimed

Koprowski's vaccine was cultivated using the kidneys of monkeys that had suffered from the Simian Immunodeficiency Virus (SIV), and that this vaccine was distributed to millions of children in the Congo in the 1960s, starting the spread of HIV.

Two elements fuelled Tom Curtis's argument: first, that molecular genetic analysis showed the HIV-1 and HIV-2 viruses were descended from the simian virus, SIV, but no one knew how the virus had made the jump from monkey to man. The other element was the discovery in 1960 that batches of Salk and Sabin's vaccines were contaminated, through infected kidney cells, with another monkey virus, SV40. SV40 was a DNA virus shown to kill cells in culture and cause cancer when injected into hamsters. Follow-up studies showed that those who received the SV40 contaminated vaccines were not at greater risk of cancer but at the time the US government covered up the results, barring Dr Bernice Eddy from publishing a study on a cancer-inducing virus (SV40) in Rhesus monkeys. While the consequences of SV40 remained unknown, an incident a few years later proved the seriousness of transmission of monkey viruses into humans when, in 1967, an unknown virus crossed from monkeys into humans in a laboratory in Marburg, Germany, causing thirty-one cases of haemorrhagic fever and seven deaths.

By 1992 the stage was set for Tom Curtis's suggestion that HIV had crossed into humans not through eating infected meat, or via monkey bites, but from infected kidney cells turned into polio vaccines at the Wistar Institute in Philadelphia and distributed by Koprowski. Curtis picked up the scent after discovering that two San Francisco scientists, Baine Elswood and Raphael Stricker, were working on a paper entitled 'Polio vaccines and the origins of AIDS'. Despite the paper still being under review, Curtis published his story on 9 March 1992. Although the story was ridiculed as having 'not one picogramme of evidence'[11] the response from Koprowski and the Wistar Institute did not engender much confidence. Koprowski, by then an old man, said that he could not remember where the monkey kidneys had come from. After a six-month investigation, the Wistar Institute concluded that there was an 'extremely low' probability that the vaccine had introduced HIV into humankind but that, as they were unable to be certain about the origin of the monkey kidneys, the presence of SIV could not be discounted.

Although Koprowski successfully sued *Rolling Stone* for libel, the author and journalist Edward Hopper picked up the trail left by Tom Curtis. In 1999, he published *The River: A Journey Back to the Source of HIV and AIDS*. Hopper was well known for writing about HIV and had undertaken research for many years in Africa. His new hypothesis was that the Wistar Laboratories had used chimpanzees' kidneys infected with SIV, rather than kidneys from Rhesus monkeys. The dispute came to a head on 11 September 2000, when Hopper, Koprowski and other scientists were called to give evidence at a two-day meeting, convened by the Royal Society in London, on the 'Origins of HIV and the AIDS epidemics'. At the meeting, Hopper threw in a wildcard, suggesting that the vaccine had been made in local veterinary laboratories in Africa, rather than in the Wistar laboratories. This was a partial response to the fact that Wistar had combed their records and proved that chimpanzee kidneys had never been part of their production line. Wistar also showed that samples of Koprowski's OPV contained no chimpanzee DNA, only monkey DNA. In addition, evidence showed that SIV could not survive in Koprowksi's vaccine.

Later research showed that the transition between SIV and HIV probably occurred decades before Koprowski's vaccine trials of the late 1950s and early 1960s, and that the strain of HIV carried into the US originated in Haiti, not Africa. The fact that Curtis's and Hopper's allegations were taken so seriously, however, points to a staggering level of mistrust in the medical profession and drugs companies. Attitudes have changed dramatically since the halcyon days of the 1950s, when doctors were father figures, who made decisions about people's health without explanation and drugs companies produced the life-saving remedies – like the polio vaccine – that the public clamoured for. Where, and why, was that trust broken?

Our bodies – our inoculations?

'People who have vaccine doubts or vaccination hesitations are from across the political spectrum,' historian Elena Conis said:

> They do not fit neatly into any one political
> camp or the other. There's been a long trad-
> ition of vaccination resistance coming from a
> libertarian ideology in the US and therefore
> it's been associated with the right wing.
> But you see something really different in
> recent decades. You see vaccination resistance
> coming from a whole host of different ideolo-
> gies. Libertarianism is part of it, conservatism
> is part of it, religious beliefs are part of it but
> so are liberal ideas about the body and nat-
> ural healing and the belief in nature as a
> better healer than organized medicine or the
> pharmaceutical industry.[12]

In her book, *Vaccine Nation,* Conis describes the development of vaccination programmes in the USA, starting with President

Kennedy's Vaccination Assistance Act of 1962, and discusses how those efforts were affected by the social movements of the 1960s and 1970s, including feminism and environmentalism.

Vaccine popularity reached new heights in the USA at the end of World War Two. In the 1950s, parents clamoured for the security vaccines such as polio brought. As a whole, the nation seemed to believe that vaccines represented safety at a time of Cold War tension, not to mention trumpeting American technical superiority. Kennedy's announcement of the Vaccination Assistance Act in his 1962 State of the Union address took most by surprise but its aim of ensuring that all Americans had equal access to medical care, regardless of their ability to pay, was widely lauded. Later considered one of the most successful prevention programmes in public health, the Act meant the US federal government took an increasing role in approving, paying for and delivering vaccines for the remainder of the decade. Between 1962 and 1964 as many as fifty million children and adults were vaccinated against polio, while seven million children were vaccinated against diphtheria, whooping cough and tetanus, reducing cases dramatically. Although statistics showed that some communities (typically the wealthy middle class suburbs) had a far higher uptake of vaccines than poorer urban communities, it seemed that a new era in vaccination had dawned; a golden era, in which 'Government scientists, triumphant over polio and smallpox, considered how the country might deploy vaccines against what they called the "milder" diseases,' said Elena Conis. With an effective vaccine on the market in 1963, the time seemed ripe to tackle measles.

Twenty million doses of measles vaccine were distributed but doctors noted that it met with little of the enthusiasm shown for the polio vaccine. Although the surgeon general reported that measles killed more children than any other childhood disease, figures collected by the CDC showed that most of the measles vaccines given in the mid-1960s were administered in private practices in white middle-class areas. And even in those areas there existed a sense that measles was a well-known, relatively harmless, part of childhood. The medical community and drugs companies needed to persuade parents otherwise and they set about this task with vigour: 'All too often measles is a killer or the cause of mental crippling so severe that the victim survives only as a mental defective', warned the vaccine-maker Merck, for example.[13]

In 1964, the CDC concluded that only a mass vaccination against measles would effectively target the disease. That view became a confident proclamation that the disease could be eliminated in the USA by 1967. Using advertisements, films, comic strips and celebrities to spread the word, the government, the medical community and pharmaceutical companies rallied to the message that measles was a serious disease that sometimes had fatal complications; in one cartoon colouring book the red-faced, yellow-eyed fiend of measles terrorizes white children in elementary schools before a vaccine sends it scurrying back into the dark, dirty city in the distance. At the height of the campaign, between 1965 and 1968, measles cases fell from 260,000 to 22,000 but the expiration of the Vaccination Assistance Act in 1969 and the failure of the Nixon administration to fund an extension in 1970 meant that the measles eradication campaign was over, and cases soon began to rise.

Parents had never been fully convinced of the urgent need for a measles vaccine. This lack of conviction became even more apparent when a vaccine for another well-known disease – mumps – came on the market in 1967. Until Merck's development of a vaccine, mumps was considered an unimportant childhood disease, which caused inconvenience, rather than any great danger. The development of the vaccine, however, led the medical community to reframe the disease as serious, a disease that could lead to brain damage or infertility in boys (a popular and still-repeated fear, although the CDC noted that no concrete evidence for this was ever produced).

'Each new vaccine shines a spotlight on its target infection and a number of things happen,' Elena Conis explained:

> One thing that happens is that the pharmaceutical companies work hard to encourage people to see the disease differently and to use the vaccine but they don't work alone. Immunization researchers and epidemiologists also suddenly have a new tool at their disposal and so they too look again and re-evaluate and reassess the target infection.[14]

As the 1970s progressed and vaccination rates stagnated across America, doctors, public health workers and pharmaceutical companies coalesced around a single message: 'Any disease that could be prevented with a vaccine was dangerous, if not deadly, to children.'[15] Polio, diphtheria, measles, tetanus, whooping cough and mumps were grouped without any attempt to differentiate between their severities. Vaccines presented in the 1960s as great social equalizers were now presented as handy middle-class conveniences, especially for the new generation of working mothers who, it was argued, could ill-spare time off to look after children suffering from preventable childhood ailments.

Supporting immunization was like supporting motherhood and apple pie, boasted one devotee of President Jimmy Carter's childhood immunization drive; you couldn't go wrong. But he could not have been more mistaken. A decade of economic traumas, oil shortages, military defeats and demands for women's, gay and minority rights shook public faith in authority and left the 1970s' political establishment reeling. Although the pharmaceutical companies correctly understood the trend towards more mothers in the workforce, they did not foresee that the second wave of the tide of feminism, in the 1980s, would usher in a new wave of vaccine scepticism in mothers who protested that they were excluded from proper decision-making by a male-dominated medical profession that had traditionally shut women out and withheld critical information about women's bodies and the substances they put in them. The doctors of the 1960s sought to calm housewives' nerves with tranquillizers and cigarettes; could they be trusted with vaccine safety when a burgeoning environmental movement, encouraged by the success of books such as Rachel Carson's *Silent Spring*, demanded a full account of ingredients and the damage they could potentially inflict?

One of the great vaccine controversies of the 1980s surrounded the DPT combined vaccine for diphtheria, pertussis (whooping cough) and tetanus. The US news network NBC aired an investigative report, *DPT: Vaccine Roulette,* which told parents that 1 in 7,000 children suffered serious effects from the vaccine, including high fevers, crying, brain damage and death. The programme included footage of seriously disabled children intercut with interviews with male doctors who denied any risk. One health professional said on screen: 'If we told parents there was a risk of brain damage . . . there's no question what their response would be.'[16] Europe and Japan had publicized the risks of the vaccine, it was banned in Sweden and West Germany and vaccination rates in England had fallen from eighty per cent to thirty per cent in the 1970s, yet the vaccine was actively promoted by the US medical establishment.

As well as causing a flood of alarmed phone calls from parents to family doctors and local television stations, the programme brought together some of the parents who believed that their children had been harmed by the vaccine. They formed an advocacy group, named Dissatisfied Parents Together: DPT. One of the mothers, Barbara Loe Fisher, who remains a prominent vaccine-sceptic, co-wrote *DPT: A Shot in the Dark,* with Harris Coulter, which contained more devastating stories: 'These doctors and officials in government, who keep talking about the benefits and risks of this vaccine, better take fair warning,' said Janet Ciotoli, a nurse whose son Richie died after his first DPT injection. 'My baby may be just another statistic to them but he was my child and there is nothing more powerful than a mother's fight for her child.'[17]

The battle over the DPT vaccine summoned the image of an army of mothers fighting to protect their children, one of the world's most powerful archetypes. It also brought together feminists arguing for more individual control of their bodies and housewives who just wanted to look after their families; women from across the political spectrum fighting for access to information, a true sense of informed consent and action to make vaccines safer. 'The metaphor of a war between mothers and doctors is sometimes used for a conflict over vaccination', Eula Biss wrote. 'The warring parties may be characterized as ignorant mothers and educated doctors or intuitive mothers and intellectual doctors or caring mothers and heartless doctors . . . sexist stereotypes abound.'[18]

The simplicity of the struggle papered over much of the complexity. What does 'informed consent' really mean when most scientists and doctors know that a single study or report can never be understood in isolation but must be interpreted within a much larger body of thousands of reports and years of hands-on experience with patients? How can people judge the true risk of a vaccine when the original smallpox vaccine, gladly and recently taken by millions of people, contained far more 'live' material, and was therefore much riskier, than all the vaccines we take today combined?

The DTP movement of the 1980s initially took a measured view, understanding that an outbreak of whooping cough would cause thousands of deaths, while the vaccine caused just a few dozen cases of brain damage a year. Barbara Fisher argued that they were not opposed to vaccination, just in favour of safer vaccines, better information and compensation for those adversely affected. 'No parent should be put in the untenable position of having to choose between a bad vaccine and a bad disease,' Fisher stated, in a wise summary of the dilemma. In time, however, the group became more aligned with traditional libertarian thinking, placing less emphasis on federal government intervention to ensure better reporting of cases and oversee standards and arguing for more parental choice in healthcare and the ability to opt out of treatments they disagreed with.

Some wanted to opt for specific vaccines for their children. Others were influenced by ideas that had come from the environmental movement. The public had discovered that things such as cigarettes, artificial sweeteners and pesticides, which they had once been assured were safe, could do untold damage and even kill them. Why should the same not be true for vaccines, argued sceptics, when 'no one knows for sure how effective or safe immunization really is and it is unlikely that we will ever know'.[19] Introducing a vaccine into a child's body, they argued, was 'toxic', ruining harmony with nature and stopping children developing their 'natural' immunity to diseases. Vaccine sceptics became adept at studying the list of chemicals included in vaccines and identifying those they believed to be hazardous and particularly those believed linked to cancer. Of particular concern was the use of thimerosal, a mercury derivative, which critics seized on at the same time as the wider debate about mercury was growing. Although the Public Health Service and American Academy of Paediatrics determined that the risk of mercury exposure from thimerosal, while unknown, was much smaller than the risk of infectious diseases, in July 1999 they stated that drugs companies should eliminate or reduce the possible mercury content of vaccines.

Fears of 'toxic' environmental pollution, prevalent in the 1970s, became 1980s concerns about the spread of immunity disorders such as AIDS. In their turn, they developed into 1990s concerns about a link between vaccines and mental and behavioural difficulties in children. Yet although these issues received much coverage in the media, the general fall in vaccine uptake could be linked more to poverty and healthcare funding than to the concerns of vaccine-sceptical parents. After President Carter's successful vaccine initiative in the 1970s, childhood vaccination languished in the Reagan and Bush years, with the CDC complaining in 1983 that they had funding to vaccinate only half as many children as they had in 1981. By the time of the measles epidemic of 1989–91 the US immunization rate was worse than nearly every country in Latin America. Lawsuits filed against vaccine manufacturers and health-care providers by parents who believed their children were affected by the DPT vaccine meant that vaccine prices soared. Several manufacturers stopped production altogether. A vaccine shortage caused public health officials to worry about the return of epidemic disease and led the US Congress to pass the National Childhood Vaccine Injury Act (NCVIA) in 1986.

The newly elected president, Bill Clinton, pointed out in 1992 that in the Western Hemisphere, only Haiti and Bolivia had lower pre-school immunization rates than the USA. He determined to change this and children's vaccination rates reached record highs by the end of Clinton's time in office, with ninety-six per cent protected against diphtheria, tetanus and whooping cough, ninety-three per cent vaccinated against measles, mumps and rubella, ninety per cent against polio and eighty-eight per cent against chicken pox and hepatitis B. Yet the link between vaccination rates and funding remains crucial, Elena Conis argued:

> The rate of philosophical exemptions [otherwise known as religious exemptions] began to creep up and in some communities went up very dramatically. And the rate of philosophical exemptions went up as the immunization schedule got longer, which means that people who are filing philosophical exemptions may be filing because they missed one vaccine or because they don't want one particular vaccine.[20]

In the USA, under-vaccination is much more commonly found in poorer communities, possibly because parents facing a long series of visits to the doctor for different vaccines means they pick and choose, prioritizing those they can afford. Thankfully, this situation does not arise in the UK where vaccinations are free under the NHS.

Moral issues have also crept into the debate. The hepatitis B vaccine was the first to cause a stir, due to its 'marketing' to gay men and drug users in the 1980s. More recently, we have seen furore over the Human Papillomavirus (HPV) vaccine offered to teenage girls. Although the HPV vaccine (developed to prevent cervical and oropharyngeal cancer) was welcomed at first, the decision by some American states to make it mandatory for teenage girls made it the target of bitter criticism from religious groups, who believed it encouraged teenagers to have sex, as well as from feminists who questioned why girls were singled out when the HPV virus also affected boys. 'Boys don't have to get vaccinated for the same reason they don't have to wash dishes, do laundry, buy birth control or think about other people in general: girls will do it for them', wrote William Saletan in *Slate* in 2009.[21] 'In the case of HPV', Eleanor Conis said:

> There were advocacy groups who saw this vaccine as a really valuable tool and also advocated for it and there were also consumers and advocacy groups that thought the opposite way and wanted to put limitations on its use. But what all of that activity meant was more and more attention to HPV, more and more awareness of what the infection was and therefore added justifications for scientific enquiry and for public conversation about it. So it's really been a trigger.[22]

In other words, the introduction of a new vaccine had once again reshaped the public's view of a disease. (The UK has a highly successful HPV school-based vaccination programme, reaching an uptake level of 86.8% in 2011–2012, with the majority of media coverage reporting positively and accurately on the vaccine – even while the controversy over MMR still raged. Even so a 2015 study in the *Journal of Public Health* revealed that the uptake for girls from ethnic minority groups in South-West England was roughly half of the level for white British girls – demonstrating a disparity that was attributed to the 'sexual' associations of the disease).

Vaccine scepticism, or outright hostility, left the medical profession confused and despairing. A group of puzzled scientists fulminated in the WHO *Bulletin*:

> Vaccination has greatly reduced the burden of infectious diseases. Only clean water, also considered to be a basic human right, performs better. Paradoxically, a vociferous anti-vaccine lobby thrives today in spite of the undeniable success of vaccination programmes against formerly fearsome diseases that are now rare in developed countries.

They mournfully concluded that 'putative vaccine safety issues are commonly reported while reviews of vaccine benefits are few'.[23]

The failure of reason

How should those in favour of vaccines respond to what often appears as a multi-pronged attack? The correspondents to the WHO *Bulletin* admitted:

> How one addresses the anti-vaccine movement has been a problem since the time of Jenner. The best way in the long term is to refute wrong allegations at the earliest opportunity by providing scientifically valid data. This is easier said than done . . . because the adversary in this game plays according to rules that are not generally those of science.[24]

The failure of the medical community to understand that the rules of the game are not the rules of science perhaps accounts for its biggest weakness. There's never been a serious argument scientifically on the anti-vaccine side,' said Steven Salzberg, a prominent scientist in the field of bioinformatics and computational biology, who regularly writes about vaccine issues. Salzberg says he was shocked, when working on the Influenza Genome Sequencing project and the development of the flu vaccine, to realize that many people didn't believe that vaccines worked: 'Even worse, they think vaccines are harmful.' Salzberg pointed out that for the doctors and scientists involved, 'There's no nefarious conspiracy here. People who work on vaccines are doing it because they've decided that this is a good thing to do with their careers and their lives as they try to make people healthier.'

Salzberg finds the reporting of vaccine issues and interventions by campaigning politicians frustrating:

> Science and medicine are not a topic where there's a need for balance. There are not two sides to most scientific questions. There's only one side, the side of the truth. And scientists, good scientists anyway, are trying to figure out where the truth lies. In the case of vaccines, there's only one side. The reason people try to develop them is to cure disease and prevent disease and there isn't another side of the issue.[25]

Salzberg's comments may well be true but falling back on the bald truth fails to take into account both the small number of scandals created by profit-hungry drugs companies that have tainted the well of public opinion and the power of the uncontrolled torrent of opinions and online abuse that rages across the Internet: so fast, so unverifiable, so little driven by reason or 'truth'. Depending on your view, anti-vaxxers are either people waging an irrational war on science, or people who interpret the concept of 'risk' differently from a public health official overseeing a global campaign. Elena Conis said that, in her opinion:

> A lot of thoughtful people who believe strongly in vaccines will concede the point that vaccine worries are not entirely irrational. Parents who get their children vaccinated are taking an act to protect their child but it's not a risk-free act. We know that there are scientific ways of evaluating risks but social scientists and humanists also know that the way in which we understand risks is a very subjective process.

Society accepts some risks; such as driving cars, or in America owning guns but, as Elena Conis noted:

> We think they're risks worth taking . . . and the parents who are weighing the risks of vaccinating versus not vaccinating are making a similar value-based judgement. So we can fault them for not making scientific decisions but I think that's entirely unfair because even the most scientifically-minded among us make subjective decisions about risk every day.[26]

Behavioural economists argue that vaccination rates will only rise when the debate is moved away from science and attempts to understand how people make important decisions about their lives. How do they evaluate the risk of disease, understand the concept of an immediate threat to their health compared to a distant prospect? And who do they trust? Some interesting findings have emerged from this work. For example, focusing a media message on the high rate of vaccination seems to motivate people more than a scare story about the number of unvaccinated children. A recent study in Philadelphia also suggests that making vaccinations easier and more convenient is far more enticing than a financial reward.[27]

There is no doubt that Edward Jenner's discovery prompted one of the greatest breakthroughs in human history. Vaccines save 2.5 million lives around the world every year, although 1.5 million children still die from diseases that could be prevented by vaccination. Yet, although more children than ever are vaccinated, the debate about vaccination remains just as contentious as it was when Edward Jenner gave the first smallpox jab. Andrew Wakefield's allegations about the safety of the MMR vaccine were debunked but they have endured, and still cloud the debate on vaccine safety. In the era of social media, truth, authority and information are hard to judge and vaccine scepticism seems likely to flourish rather than diminish. As a result, vaccines may remain the most successful and cost-effective public health story never told.

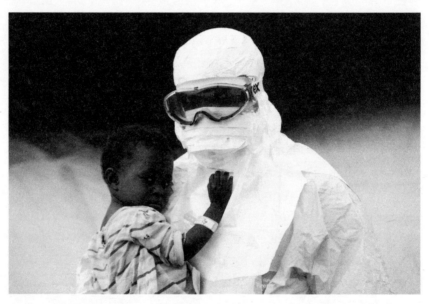

The Ebola outbreak dominated global headlines and decimated parts of West Africa but the urgent campaign to treat and contain the epidemic benefited enormously from the polio vaccination infrastructure in the region.

8

EBOLA: AN AVOIDABLE CRISIS

When that epidemic has been stopped . . . [the] Ebola virus will not be
gone. It will only be hiding again. It will recede into its reservoir host,
somewhere amid the forest and await its next opportunity.
David Quammen[1]

When Patrick Sawyer, a forty-year-old Liberian civil servant died of Ebola in a hospital in Lagos, global health workers in Nigeria were hard at work on the final stages of polio elimination. The day before Sawyer's death, 24 July 2014, the worried parents of a sixteen-month-old toddler in Kano state had discovered their child had symptoms of poliomyelitis. Given the terrifying spread of Ebola and the prospect of it taking hold in West Africa's most populous country, no one in the media paid much attention to the polio diagnosis but, although the two viruses are very different, the efforts to tackle polio and Ebola were intricately connected. While the polio eradication campaign provided a crucial blueprint for containing, monitoring and treating outbreaks of Ebola, the emergence of new diseases imperilled the success of long-standing eradication campaigns.

For Dr Failsal Shuaib, the polio diagnosis of 24 July was a tough moment. The polio eradication campaign was within a whisker of the phenomenal breakthrough to zero transmissions in Africa but this child was paralysed, because she had missed some in the series of immunizations so important in tackling the virus in areas of poor sanitation. 'I grew up seeing people crippled by polio left along the sides of the road,' Shuaib said. 'Eradication is within reach but the gains we have made are fragile and we must not let the fight to tackle Ebola set the clock back.'

Nevertheless, Nigeria's ability to tackle Ebola was directly helped by the work of Shuaib and his team. Shuaib grew up in northern Nigeria, qualified as a doctor and took a position in public health for the WHO, becoming deputy incident manager at Nigeria's polio Emergency Operations Centre (EOC). When Nigeria's first Ebola case was diagnosed, he was immediately transferred to lead an Ebola EOC.

'The Ebola virus coming to Nigeria was something new and it's been tough,' Shuaib said:

> Working on the polio programme is very demanding but having to manage an Ebola outbreak of this magnitude has meant living with the possibility that if you don't get it right it could have potentially overwhelmingly catastrophic consequences for the whole of West Africa. When you see your colleagues dying around you, it wears you down and saps all your faculties.[2]

Nonetheless, by implementing the techniques that had proved so successful against polio – setting up an EOC to ensure a quick and coordinated response, and establishing a rapid house-to-house surveillance system – Shuaib and his colleagues brought the Ebola epidemic under control. The Ebola outbreak demonstrated the crucial lessons that eradicating a disease of the past could offer for containing the diseases of the future but it also threatened to wipe out many of the gains made.

At the WHO in Geneva, Hamid Jafari said, the polio eradication campaign was affected by the Ebola outbreak in terms of 'active planning . . . being lucky and going through periods of being really weakened in our capacity'. The terrible nature of the Ebola outbreak called for an emergency, all-hands-on-deck, approach, pulling in workers and strategists from all diseases and specialities, Jafari said:

> Polio supported large chunks of the Ebola response. We also got lucky in the sense that Nigeria managed to rapidly control the Ebola outbreak, mainly through the use of the polio staff. They were well trained and knew how to move in the field, do surveillance and train people in tracing those who had come into contact with the disease.[3]

The campaign was also lucky, he thought:

> Because polio didn't get into Ebola-affected countries during that time period. It was a credit to Nigeria that it was in an aggressive polio programme control phase when the Ebola outbreak was happening, so we didn't see a spread of polio. We would have been in

a very tough situation if Ebola and polio were
co-travelling in those Ebola-affected countries because a lot of the health capacity was
lost to deal with any other outbreaks.[4]

Despite polio staff being diverted to deal with the Ebola crisis, and
delays in the polio programme, Nigeria continued to show strong
progress against polio. The effect for other diseases was tougher.
The director of the Johns Hopkins Malaria Research Institute,
Peter Agre, said:

There was so much alarm caused by the
Ebola crisis in West Africa that people stopped
going to clinic for routine malaria treatments
or disease prevention in pregnancy. So in
West Africa there were probably many, many
more deaths from malaria due to the panic of
Ebola than from the Ebola itself.[5]

Research commissioned by the Global Fund to Fight Aids,
Tuberculosis and Malaria and the US President's Malaria Initiative
concluded that in Guinea as many as 74,000 cases of malaria went
untreated during the Ebola outbreak, as many people with malaria
were afraid of catching Ebola at a treatment centre or feared they
would be sent to an Ebola clinic by mistake.

The stress the Ebola outbreak placed on Guinea's weakened
health system resulted in a drop in vaccination coverage overall
and threatened to derail what had been Africa's most successful
health campaign, for meningitis A. Angela Hwang, an expert in
vaccine delivery at the Gates Foundation, said:

Ebola struck and that delayed routine imple-
mentation in Ghana and slightly delayed
the second tranche of countries. A lot of the
people who are part of managing epidemic
meningitis were pulled part- or full-time into
the Ebola work and they kept the meningitis
work going on, on the side during nights and
weekends.[6]

'It's a river – in Africa'

Forty years after its first recorded outbreak, Ebola brought healthcare systems in West Africa to their knees, imperilled other disease
programmes and struck terror across the globe. But the disease
remained mysterious, with a suspected but unproven, animal reservoir in bats, and long stretches with no human transmission. The
microbiologist and director of the London School of Hygiene and
Tropical Medicine, Peter Piot, was then a junior researcher in a
laboratory in Antwerp that was certified to deal with viruses
including yellow fever. He remembered the very first outbreak:

> On the last Tuesday in September 1976 my
> boss at the microbiology lab was alerted that
> a special package was on its way to us from
> Zaire. It was flying in from Kinshasa; samples
> of blood from an unusual epidemic that
> seemed to be stirring in the distant Equateur
> region, along the river Congo.[7]

Several nuns were known to have died of the as yet un-
determined disease, despite their yellow fever vaccinations being
up to date.

The next day, the flask arrived. 'A cheap plastic thermos flask, shiny and blue,' Piot recalled. With their hands protected only by latex gloves the team in Antwerp opened it. 'Unscrewing the thermos we found a soup of half-melted ice; it was clear that sub-zero temperatures had not been constantly maintained. And the thermos had taken a few knocks, too.' One of the test tubes was broken, 'its lethal content now mixed up with the iced water' and five millilitres of blood from a dead Flemish nun. Samples sent to Porton Down in the UK and the CDC in America established that while the virus was similar to the Marburg virus, it was not Marburg but a new and completely unknown virus. Much to his excitement, Piot was seconded to travel to Yambuko, the mission station that appeared to be the centre of an epidemic that had been raging for three weeks and claimed two hundred lives. 'There's a kind of excitement that takes over when real discovery is at hand,' Piot said. 'I was facing a total unknown: a new virus, a new continent. Even the insects were monstrously unfamiliar . . . But I had no room for fear or worry, I felt *alive*.'[8]

Alive in the face of death, terror and uncertainty. Piot described the first time he entered an African home, wearing gloves, gown, motorcycle goggles and face mask, to meet the sick couple who lay inside, sprawled on raffia mats: 'Flies settled insistently on the black, crusted blood around their mouths, noses and eyes. Both had dark blotches on their torsos and their eyes were darkly bloodshot. They had barely the strength to move.' The man began vomiting blood painfully and spottily. As Piot and his colleague took blood samples from the woman, the man died.

Piot laboured on in conditions of varying hardship and uncertainty, once narrowly avoiding death by refusing to travel in a doomed helicopter flown by drunken pilots 'who caused havoc all over the region, blowing the roofs off huts and sleeping with an apparently endless loop of girls'. Although episodes like this drove him to the edge of nervous breakdown, Piot and his colleagues monitored and charted the spread of the epidemic, sketching out an epidemiological curve that showed, day by day, when people became ill and predicted when cases might subside.

Late one night, while swigging bourbon from a flask, the team debated what the new virus should be called. The first, and simplest, suggestion, to call it the 'Yambuku' virus, was dismissed: too much stigma for one small town. Calling it after a river seemed a suitable alternative but the most famous river in Zaire, the Congo River, was already the name of a horrible disease. The team huddled around a small map of the vast country and settled on the name of what seemed to be the closest river to Yambuku, whose name translated into English as 'black river'. Piot later noted that, 'Actually there's no connection between the haemorrhagic fever and the Ebola River. Indeed, the Ebola River isn't even the closest river to the Yambuku mission. But in our entirely fatigued state, that's what we ended up calling the virus: Ebola.'[9]

Although the disease's symptoms of high fever, diarrhoea and (sometimes) bleeding were horrible, and the death rate among the infected was terrifyingly high, the disease did not seem to be airborne, making it containable by controlling person-to-person transmission. The Yambuku outbreak eventually spread to fifty-five nearby villages. There were 318 cases of the disease and 280 people died. Yambuku, however, was not the site of the first Ebola outbreak; that occurred further north, in a cotton factory in Nzara, Sudan; it claimed 151 lives. Further outbreaks across East and Central Africa over the next four decades provided more clues about Ebola's origin and evolution, ultimately revealing that Ebola is not one virus but five and that it had existed in the region for years before the first case was reported in the 1970s.

The International Commission flown in to oversee the Yambuku outbreak reported that, 'No more dramatic or potentially explosive epidemic of a new acute viral disease has occurred in the world for the past thirty years.'[10] Yet no more outbreaks were reported for twenty years, until the Tai Forest strain and the Zaire strain reappeared in a series of outbreaks in 1994 in Gabon and Zaire. The disappearance of the virus for years at a time made it particularly hard for virologists to track and to identify the animal reservoir. As David Quammen wrote, they might 'Look for Ebola anywhere, in any creature of any species, in any African forest but those are big haystacks and the viral needle is small.'[11]

The outbreaks differed in scale; some caused a few dozen cases and claimed a handful of lives but others were explosively contagious. When a farmer and charcoal collector who lived on the edge of a forest in Kikwit, near Kinshasa, contracted Ebola, it ripped through two hospitals; 315 people caught the disease and 254 died. The fear, as when the 2014 outbreak reached Lagos, was that the disease would reach a major metropolis, where it might kill tens of thousands.

Although much remained unknown, certain elements appeared consistent: all the outbreaks began near the forest, where villagers came into contact with contaminated animals and they were spread when people came into bodily contact with others who had the disease, such as in hospital, at a traditional healer or when cleaning and preparing the body of a victim for their funeral.

By 2013, 1,590 people had died in the outbreak of Ebola. But even though the symptoms were often not as gruesome as commonly reported, Ebola was a disease of fear; a disease that 'killed with a whimper, not with a bang or a splash'.[12] The chair of the National Ebola Taskforce, Dr Sam Okware, who oversaw the response to an outbreak in Bundibugyo, Uganda in 2007, said the terror of the local population had been the most difficult thing to contain: 'There was a new epidemic – of panic.'[13] People living in the remote villages affected by the disease were shunned. The local banks closed, people wouldn't accept their money and even those fortunate enough to recover found that their homes were burned down upon their return.

And still scientists and doctors remained in the dark about key aspects of the disease. Why, in one instance, had only about half the people in a family been infected? Why were some villages struck and not others? Was the animal reservoir bats, as seemed the most likely conclusion?

'An outbreak like no other'

Then came the outbreak of 2014. Writing in the *New England Journal of Medicine* in October that year, Peter Piot and Jeremy Farrar outlined how the response to the disease signified some of the challenges and failures that had plagued the international health community for years, and warned:

> The twenty-fifth known outbreak of the Ebola virus infection is unlike any of the previous epidemics. It has already killed . . . more than all previous epidemics combined; it's affecting virtually the entire territory of three countries, involving rural areas, major urban centres and capital cities; it has been going on for almost a year; and it is occurring in West Africa, where no Ebola outbreak has previously occurred. Above all, the epidemic seems out of control and has evolved into a major humanitarian crisis that has finally mobilized the world.[14]

'Patient Zero' of the outbreak was Emile Ouamouno, a two-year-old boy from the Guinean village of Meliandou, who fell ill in December 2013. A few days later Emile died, followed by his three-year-old sister Philomène and their mother, Sia, who was pregnant. Unwittingly, a local nurse had treated them for malaria. When Emile's grandmother became ill, she left the village to seek treatment, carrying Ebola to another town and spreading the disease. A doctor, nurse and health worker died. 'We thought it was a mysterious disease', the director of the nearby Guéckédou Hospital, Dr Kalissa N'fansoumane, told the *New York Times*.[15]

Emile's bereft father cursed the village for killing his family but two hours from the nearest city, lying deep in the forest and surrounded by palm cultivations, Emile's death in Meliandou went unheeded for twelve weeks among the wider health community. In the meantime, Ebola smouldered across porous borders between Guinea, Liberia and Sierra Leone, three countries with poor health systems, few doctors and a constant cross-border traffic of people going to market or visiting their families.

On 10 March 2014 local health officials alerted Guinea's Ministry of Health about the outbreak of deaths and illness. Doctors from the capital and the charity Médecins Sans Frontières converged on Meliandou to take blood samples. No case of Ebola had been diagnosed in West Africa since a single case in Côte d'Ivoire in 1994, where the lucky patient had recovered, but on 21 March an email sent to the WHO in Geneva by Sylvain Baize, an infectious disease specialist at the Pasteur Institute, confirmed the news: Ebola was spreading in the region, with 49 suspected cases and 29 deaths. Unlike the case in Côte d'Ivoire, which was the Tai Forest strain, the samples showed that this was the Ebola variant, arising from Gabon, the Congo and the DRC. It had travelled two thousand miles east and was mutating rapidly; the samples showed that this strain had ten years' worth of mutational differences from the DRC strain. The conclusion was that the disease had been evolving independently for ten years, but no one knew where.

Experts had no reason to believe that containing this epidemic would be more difficult than dealing with any other. In early April, Dr Pierre Rollin from the CDC arrived in Meliandou. His team worked for five and half weeks to trace and contain the disease; successfully, it seemed. The number of Ebola cases fell week by week and new patients slowed to a trickle. Doctors in the Guinean capital Conakry reported that they had not seen a case in ten days. Liberia said it had been a month. Sierra Leone had not reported a single case. 'That's it for this outbreak,' Rollin told himself, returning to Atlanta on 7 May confident it had burned itself out. However, the virus was far from finished and it seemed that the experts had missed the signs.

Compared to airborne diseases, such as influenza or SARS (Severe Acute Respiratory Syndrome), Ebola is slow-moving; so how did the 2014 outbreak smoulder undetected and then spread across West Africa, causing 28,637 people to fall ill and taking 11,315 lives?[16] The answer appears to be a combination of oversight, poor communication and slow response.

Ferocious civil wars and *coups d'état* had left the three countries worst affected by Ebola (Liberia, Guinea and Sierra Leone), with weak governance and poor health infrastructure. Charities and international health groups had pulled out during the various wars and periods of unrest and the region lacked investment in even the most basic structures that might have helped contain the disease: doctors, treatment centres and beds, not to mention a shortage of gloves, masks, gowns and bleach. Liberia was reported to have fewer than 250 doctors for 4 million people. As the WHO assistant director general Dr Keiji Fukudu admitted: 'This kind of outbreak would not have happened in an area with stronger health systems.'

Moreover, local people did not trust their governments to act in their best interests. Health workers found their way to villages barred by felled trees or angry mobs. Frightened families hid ill relatives, rather than send them to treatment centres that they believed might kill them. And as with other disease campaigns and epidemic responses, international organizations moved slowly, in the face of internal politics and funding issues. As David Quammen wrote:

> The events in West Africa (so far) tell us not just about the ugly facts of Ebola's transmissibility and lethality; they tell us also about the ugly facts of poverty, inadequate health care, political dysfunction and desperation in three West African countries and of neglectful disregard of those circumstances over time by the international community.[17]

An investigation by the *New York Times* laid out a series of failings attributed to weak regional government and underfunding and WHO internal politics. Although Head of the WHO, Dr Margaret Chan, admitted that 'Hindsight is always better', and that everyone, including the national governments involved, had 'underestimated this unprecedented, unusual outbreak', it appeared that breakdowns in the monitoring and reporting of cases over two months in the summer of 2014 allowed the disease to spread significantly.

When Pierre Rollin returned to the CDC in Atlanta he was confident that the disease was under control. Not only were cases falling overall but Sierra Leone had not reported a single case. Later, it emerged that Ebola had crossed the border into Sierra Leone and spread for two months before the authorities were alerted. A traditional healer caught Ebola from patients in Guinea and infected people in Sierra Leone. Some of those infected travelled back to Guinea and infected people there, restarting the outbreak. The *New York Times* discovered that cases of Ebola had been documented by the WHO and Guinean public health workers as early as March but details did not reach the authorities in Sierra Leone. Officials blamed each other, blamed language difficulties, or claimed never to have received documents. It was a devastating failure: Sierra Leone eventually reported more Ebola cases than any other country.

Cases soon emerged in Liberia, including people living in the capital, Monrovia. By July, Médecins Sans Frontières, which in March had first raised the alarm about the full scale of the outbreak, stated that Ebola was 'out of control'. A month later, on 8 August, the WHO declared that the outbreak was at the highest threat level and a 'public health emergency of international concern'. It later transpired that in the months between March and August the WHO was fraught with internal conflict. In the early stages of the epidemic, the WHO delegated leadership to its African regional office in Congo-Brazzaville, whose budget for epidemic preparedness had been halved over the previous five years; nine of its twelve emergency response specialists had been laid off. The Africa office appeared to be more or less independent of the WHO Head Office, with constant jockeying for position and authority between it and the WHO in Geneva. Specialists from Geneva were passed over in favour of local health officials with no experience of epidemics. According to the former WHO Executive Board member, Dr Nils Daulaire, the Africa office was staffed by appointees who were 'not the cream of the crop'. It was, he added, was 'a place where politics often trumps substance'.[18]

Other international agencies, such as the CDC, complained they were hampered by needless bureaucracy. The CDC's director, Dr Thomas Frieden, protested that his team, headed by Pierre Rollin, was subject to unnecessary demands, including a requirement to produce CVs. Frieden noted that he had heard that the Africa regional staff wanted to prove that they could handle the epidemic and it was unclear which branch of the WHO was in charge. 'People shouldn't die because someone's embarrassed that they can't do it themselves,' Frieden said. But that was what he believed to be happening.

At the beginning of the epidemic it was estimated that only eight per cent – a shockingly low figure – of those who had come into contact with the disease at its epicentre in Guinea had been traced. A CDC team that arrived in Conakry in April discovered just a single pair of WHO trackers, who managed to see only half of the people registered as contacts. Recruiting more workers proved difficult, as there were problems paying them (the local going rate was just over US$4 a day). The regional coordinating centre for Guinea was not set up until July and even then, it lacked the efficiency of data sharing and rapid response of the EOCs adapted from the polio campaign and used against Ebola in Nigeria.

By September, the WHO in Geneva had asserted its control over the outbreak, deploying crisis specialists rather than local country representatives. The *New York Times* claimed the lack of urgency and poor coordination could be laid only at the door of local government and the regional offices of international institutions. 'The WHO had to cut nearly $1 billion from its proposed two-year budget, which today stands at $3.98 billion'; far less than the CDC budget, the newspaper reported:

The cuts forced difficult choices. More emphasis was placed on efforts like fighting chronic global ailments, including heart disease and diabetes. The whims of donor countries, foundations and individuals also greatly influenced the WHO's agenda, with gifts, often to advance specific causes, far surpassing dues from member nations, which account for only twenty percent of its budget.[19]

At the WHO, the outbreak and emergency response team suffered deep cuts and the break-up of the epidemic and pandemic response unit into separate teams. The offices in Geneva resembled a ghost town, it was said. Senior staff issued ominous warnings: 'You have to wonder, are we making the right strategic choices?' assistant director general Dr Fukutu, mused. 'Are we ready for what's coming down the pipe?'

Perhaps a better way to phrase that question would be to ask if international agencies have ever been ready. Before the 2014 Ebola outbreak the WHO unit responsible for epidemics had only fifty-two employees, responsible for covering a range of diseases including yellow fever, bubonic plague, flu and cholera, and just one Ebola specialist. The efforts of the tiny smallpox team, and the collapse of the yellow fever and malaria campaigns in the 1960s, demonstrated that the WHO had always been underfunded. Disease campaigns jostled for attention from national governments and donors and were subject to competing trends and differing opinions about health agendas. Even the institution's biggest admirers admitted it could be slow and bureaucratic, and short of forming a world government, always would be. Short-term budget cuts cannot be held solely to blame; the bigger issue of whether the

WHO was fit for purpose and what funding and decision-making were needed to make it fit, has never been resolved.

In West Africa, the aftermath of Ebola is profound. Survivors feel stigmatized and unwanted, living lonely lives, in physical pain. They feel guilt that they are still alive but worry that they will be forgotten. 'I don't take medicine because I don't have the money,' Sullman Jalloh, who has returned to work as a hospital cleaner in Freetown, Sierra Leone, told the *Guardian*. Ebola has left him with serious back pain but he cannot afford painkillers. Jalloh has little faith that either global NGOs or his government will come to his aid: 'I hope that God will help me,' he said.[20]

The Ebola outbreak undoubtedly shone a harsh spotlight on the devastation caused when weak health systems buckle in the face of a major epidemic. However, one of the few positive outcomes was that it highlighted what single-disease campaigns such as polio eradication can offer, if they are tightly woven into the health systems and needs of the countries they serve. Beyond the summit of eradicating a disease, there is always a further mountain to climb.

The legacy of the global effort to beat polio has been passed to other diseases. This young woman is receiving MenAfriVac, a vaccine developed specifically to treat meningitis A in the African 'meningitis belt'.

9

THE LEGACY

The experience from smallpox eradication in 1980 demonstrates that the assets from a global health initiative can disappear very quickly.

'The Global Polio Eradication Initiative'[1]

Ashok Maloo sits in his small office at WHO in Geneva. It is a sunny day; beyond his window the WHO campus rolls invitingly down to the lake but Maloo has the blinds down and is pointing at a photograph of a Guinea worm, a parasitic nematode roundworm, being pulled out of a human body and wound around a thin wooden stick:

> It is medically advisable to roll it out. That's the only way you can take it out. There is no vaccine, there is no medication, there is no preventive treatment as such; at the most what you can give is painkillers to a person that's got it. You have to wait for the worm to emerge and prevent the person from

reaching water sources so that you can really wind it out of the body. It may take a day, it may take two days, it may take a week, to just slowly bring it out.[2]

Maloo produces another photo of a worm bursting through a human body:

When the worm is about to emerge, it creates a blister and the worm emerges from that. As soon as it emerges it causes excruciating pain, burning pain and that's what makes people rush to ponds and rivers to get relief from the burning sensation. But the moment you put your leg or the affected part in the water, the head of the worm splits. It explodes on contact with water and it releases millions of larvae into the water and that's how the disease cycle perpetuates.

Maloo and his colleague work next door to the polio team but their resources are tiny by comparison. Despite their modest surroundings, however, Guinea worm disease, dracunculiasis, may beat poliomyelitis and become the second disease to be eradicated in human history.

Diseases of the poor

'This disease is described in the Bible,' Maloo said. 'It's in Roman literature, it has affected armies throughout history, and we will probably be very much privileged in this current generation to see the end of this disease that has existed for thousands of years.' Like the campaign to eradicate polio, the work of wiping out Guinea worm was first proposed in 1981. Regulations about safe drinking water and sanitation led the WHO and CDC to formulate a strategy, boosted by a commitment by the former American president Jimmy Carter and the Carter Center, to make eradication of the disease one of their primary aims.

Carter became involved in Guinea worm eradication after visiting Ghana. There, he saw a woman cradling her breast. Thinking she was breastfeeding, he went over to meet the child. When he spoke to the woman, he realized she was in excruciating pain, with a worm emerging from her breast, just one of eleven surfacing from her body.[3] Although President Carter had spent some months receiving treatment for cancer, in August 2015 he said: 'I would like the last Guinea worm to die before I do.'[4] That is now a distinct possibility.

The disease is carried by tiny water flies that live in stagnant water. When people drink the water the worm larvae enter the body, where they mate in the stomach and gradually (taking up to a year) mature into a worm that can be a metre long. When mature, the worm burrows through the body and eventually bursts through the skin. For those infected there is neither vaccine nor medicine.

Several factors make Guinea worm a strong candidate for eradication: it is easy to diagnose (people can see the worm and recognize it), there is limited geographical transmission, the disease is seasonal in nature, is not airborne and there is no animal reservoir. Public health workers have set up a surveillance and containment strategy in villages where Guinea worm is endemic, creating reporting tools and preventing anyone with an emerging worm from entering water. As a result, Guinea worm disease fell from an estimated 3.5 million cases across twenty-one countries in

Africa and Asia in 1986, to just twenty-two cases in 2015, in only Chad, Ethiopia, Mali and South Sudan.

Averting up to eighty million cases of this debilitating and painful disease makes the eradication of the Guinea worm an astonishingly successful programme by any standards. Even more notable is the fact that this has been achieved despite dracunculiasis being one of a handful of Neglected Tropical Diseases (NTDs), known to cause disability and pain in millions of people around the world. These are the diseases of the poor. The WHO's Department for the Control of Neglected Tropical Diseases covers twenty such conditions, including Guinea worm disease and onchocerciasis (river blindness), another strong candidate for eradication. Yet although a billion people suffer from one of these debilitating, and sometimes fatal diseases, budgets for prevention or treatment have remained small. They are often overlooked in the drug development pipeline; NTDs account for the vast majority (ninety per cent) of the global disease burden but attract only ten per cent of health-related research funding; less than one per cent of new drugs registered between 1975 and 2004 were for treatment of NTDs. Ashok Maloo explained the scale of the challenge:

> At HQ, we have a coordinator, a technical officer and two assistants. It's a very small team and we coordinate with the national Guinea worm eradication programmes of each country wherever the disease is endemic at the moment. Sometimes we have donated vehicles and bicycles for volunteers, because much relies on the help we get from volunteers. South Sudan is a vast country. Ethiopia is huge.

In some of these countries, one province can be the size of Switzerland. 'To be able to employ people to do that job is impossible,' Maloo said:

> You need to be able to go to the village and enlist the support of the village chief, to try and canvass people to become your brand ambassador or your volunteer to go door by door, just like they do for polio. This is where we collaborate and we coordinate activities, for example polio eradication activities and vitamin health days, as they are called, where families meet, and UNICEF will be there and the polio team will be there. It cuts across different activities.[5]

Guinea worm eradication might be momentous in itself but it rates as only a minor part of the jigsaw of interlocking diseases and factors that keeps people poor and sick. In the bigger picture, transmissible diseases account for only a small fraction of a growing health burden that includes maternal mortality and non-communicable diseases such as cancer.

Creating a global control and elimination programme for NTDs would cost an estimated $1 billion, according to Peter Hotez at the Global Network for Neglected Tropical Diseases; although such a strategy is still a work in progress, the Gates Foundation and other aid agencies are putting together a more coordinated and large-scale approach. Ultimately, the legacy of a large single disease eradication campaign, such as polio, driven by first world ambitions, will be entwined with success in eradicating diseases such as Guinea worm and establishing a basic health infrastructure that meets the needs of the people it serves.

Killing the goose that lays the golden egg

The biggest question of all is, can the polio campaign build such a legacy? Over the decades the polio campaign has amassed huge resources and huge experience but earlier crusades have demonstrated how the end of a battle can see hard-learned lessons and skills disappear.

Tucked away in the largely silent, tastefully international, open-plan office of the polio programme at the Gates Foundation, Lea Hegg devotes half of her time to planning the legacy of the polio eradication campaign. She works with the Global Polio Eradication Initiative to develop guidelines and a planning process for countries to evaluate what resources they have and how they could be used in other areas. In February 2016, Hegg, together with Stephen Cochi, Anjali Kaur, Carol Pandak and Hamid Jafari published a paper, 'The Global Polio Eradication Initiative: Progress, Lessons Learned and Polio Legacy Transition Planning', which stated:

> The global health community has an obligation to ensure that these lessons and the knowledge generated from the initiative's experience are shared and contribute to real, sustained changes in our approach to global health'.[6]

The initiative is operating under a strategic plan that now extends from 2013 to 2019 with a cumulative budget estimate of $7 billion. The large workforce of more than 30,000 health workers – largely volunteers – are paid for by the partner organizations.

Pointedly, the paper continues: 'It is important to understand that the initiative will cease to exist at the conclusion of this period, so it is essential to accelerate planning for the transition from a focus on polio to other goals; a process known as polio legacy transition planning.'

'This is really pioneering. This whole transition, and how we think about it, is unchartered,' Lea Hegg said. 'A lot has been invested and a lot has been developed: expertise, infrastructure, human resources. There are over thirty thousand people that are directly employed by the polio programme and that doesn't count vaccinators and hundreds of thousands of other people.' Hegg speaks rapidly; her pace underlying the urgent and unenviably enormous task ahead:

> The programme has responsibility to ramp those things down from the perspective of polio . . . But beyond that it's hard for the polio programme to define what the legacy of its work really is. What you have to do is gener- ate demand from other programmes and find out what they want to take on.[7]

What seems like a 'neat and tidy' process on paper is difficult and demanding in practice, Hegg said: 'Because the polio programme has had its head down on eradication for so long, it hasn't really developed the expertise in reaching out to other diseases or donors or governments for that long-term planning.'

Currently Hegg is working on a timescale for transition that relies on the WHO certifying the eradication of polio in 2019, and a budget that has been developed for those nail-bitingly short few years. Certain functions will need to remain in place after polio has been eradicated: surveillance, a laboratory network, an emergency response capacity and a routine immunization programme. But what will support the programmes developed to work alongside polio eradication, such as the social network that promotes breast-feeding and hand-washing, or the tens of thousands of health workers who might find themselves out of a job? What will sustain the networks built with religious leaders and local communities?

'How do you move those resources smoothly on to something else and how can you tell whether they're valuable or we should get rid of them?' Hegg asked. 'We want to avoid the sense that the polio programme is just trying to hoist people off to other programmes. It's not that we want to take the infrastructure that was developed and just pivot it to something else, unless there really is demand for that.'

Negotiations must begin between the polio team partners, the governments of the countries involved and the donors. In countries such as India, that have large-scale, functioning health systems, it is possible to imagine transferring resources to the government but less so in places such as South Sudan, where health services are provided by NGOs. 'There are polio programme resources, particularly surveillance infrastructure, in countries where there's some level of government infrastructure that can absorb it or can plan to absorb it; countries like Indonesia, Bangladesh or Kenya and Ethiopia,' Hegg pointed out:

> But the nature of the polio programme is that the largest investments have been in those countries where they had the weakest public health system. We've put a lot of resources into countries like South Sudan, Chad, DRC and Somalia, where there isn't that much infrastructure. In those cases we need to really need to think creatively about what happens.

The big question is about sustainability: who will fund the ongoing health work that emerges from the polio eradication campaign? Without the coherence offered by the coordination of the polio programme, funding will surely drift into other areas or dry up altogether. Even if resources can be found, Hegg pointed out that it is not feasible to take polio workers, currently paid five or six times the local going rate, 'and plonk them into the Ministry of Health and find out a funding stream, because it's just not sustainable that way either'.

What to do with the thousands of grassroots health workers currently devoted to polio is one of the thorniest issues under discussion. Some NGO workers speculate that polio eradication will never succeed, because the workers have little incentive to kill the goose the lays the golden egg. Ivan Gayton from MSF said he believes this makes polio eradication impossible:

> I think they're toast. I once interviewed a woman from the polio campaign who wanted to come and work for us and she said that in Pakistan the polio campaign is such an incredible resource for the Ministry of Health that they have to make a good show of it but they can't afford to eradicate polio because it's a lifeline for tens of thousands of people who make a living out of it. So they've got to do a good job but they can't actually eradicate polio.[8]

Immunization for all?
If polio resources were to move into another area, Hegg said she sees the clearest demand for routine immunization work:

The UNICEF staff who work on the cold chain funded by polio are working on cold chain for the whole routine immunization system. Countries understand how valuable that is. They certainly recognize that the quality of routine immunization in a lot of these countries will slide back if those resources are ramped down and there's no real plan in place to protect those activities.

Hegg points to the example of Nigeria, where the polio campaign has successfully been brought together with anti-malaria work and routine immunization. The Gates Foundation claims to have overcome some of the problems other agencies have faced by sitting down with state governments and working out locally owned plans.

For many NGOs and aid groups, transferring money straight into government bank accounts, as the Gates Foundation does, is a risky, almost unthinkable, concept. 'In Nigeria, we are putting our money in a government bank account and we're asking for reports back from them. And that is shocking to most other aid organizations. They could never do that. But that's the key to this, because then the government has it in their account,' Hegg said. 'And they have to be personally accountable to Bill Gates and Aliko Dangote.' Whether a plan that relies on being personally held to account by Bill Gates in a twice-yearly teleconference is feasible the world over remains to be seen. Nigeria is riven by corruption; even newly delivered equipment often doesn't work. An article in the journal *Tropical Medicine and Health* in 2014 stated:

> Over the years Nigeria has received huge quantities of cold chain equipment. Despite

> this support, much of the cold chain appears
> to be beyond repair. This is partly due to the
> focus on polio eradication, which uses freez-
> ers. In one store, only one of the three cold
> rooms was working . . . At the state level, the
> cold stores are poorly equipped and badly
> managed. More than half of the refrigeration
> equipment is either broken or worn out.[9]

The challenges of legacy planning are among the problems, Hegg said, that the Gates Foundation seeks to overcome. And of course, they believe that they can do it. And that they are doing it.

The kid at the end of the road

The debate about eradication pits the benefits of 'vertical' single disease campaigns against 'horizontal' health service provision for all. But some believe the polio campaign is creating a new 'diag-onal' model, contributing towards building better health services while intersecting with other disease control and elimination efforts, such as NTDs or meningitis A, the target of a hugely successful campaign across the 'meningitis belt'; a broad swathe of sub-Saharan Africa, stretching from coast to coast, where meningi-tis is endemic and epidemics are frequent.

If and when polio is eradicated, and attention swiftly moves elsewhere, a host of diseases are lined up to be the next vaccination success. Thanks to a vaccine – MenAfriVac – developed specifically for the region and made available to twenty-six African countries at a tiny fraction of the cost of other meningitis vaccines, this now includes meningitis A. Angela Hwang at the Gates Foundation described how MenAfriVac became one of global health's great success stories:

MenAfriVac was championed by Marc LaForce, who became director of the Meningitis Vaccine Project for PATH and the WHO . . . He's a real powerhouse . . . a bulldog, a visionary in this space and he understood the issue of epidemic meningitis in sub-Saharan Africa and the opportunity to make a vaccine that was tailor-made for that region and could be used to actually prevent epidemics as opposed to mop-up afterwards. That was the genesis of the Meningitis Vaccine Project.[10]

LaForce approached the big pharmaceutical companies that manufacture meningitis vaccines for the rich countries of the world and asked them to make a 'stripped-down' version that addressed the specific needs of the people who needed it most. But 'big pharma' said no. 'It was not an attractive market for them,' Angela Hwang explained. 'Companies like GSK didn't have the perspective that there's actually shareholder value to be found in serving global markets.' Instead, LaForce worked with the Serum Institute, a private Indian company that manufactured low-cost vaccines for developing countries and together, they developed a 'conjugate' vaccine (one that is designed to elicit a strong response). Targeted at 1 to 29-year-olds, the single dose vaccine was launched in 2009, starting in the highest-risk countries and methodically working across the meningitis belt, applying for GAVI support for the vaccine doses and for some operating costs.

'There were huge campaigns that would target substantial portions of their population in like a week or two weeks!' Hwang said. 'There are some countries that just went end to end and had massive campaigns. It was a party! And [they] achieved coverage rates of a hundred per cent. There was huge demand for the vaccine.'[11]

The goal of the Meningitis Vaccine Project is very carefully couched as: 'Eliminating serogroup A meningococcal meningitis epidemics as a public health problem in Africa.' 'Now, that's really super public healthy,' Hwang laughed. 'We have to put on these qualifiers because we can't over-promise. We want to be very, very specific and keep it to something we can accomplish.' She toys with the idea of talking about elimination but errs on the side of caution. Nonetheless, the success of MenAfriVac has been spectacular; the WHO estimated that 250 million people in the meningitis belt will receive the vaccine over the next eight years.

The GPEI legacy planning states that: 'The knowledge and experience garnered from polio eradication efforts provide important lessons on how to reach every child, including the most underserved, migrants, nomads, people living in conflict zones and others marginalized by circumstances that prevent or impede access to health services.'[12] Angela Hwang agreed that the meningitis A work has benefited hugely from other disease campaigns:

> What we get from efforts such as the measles and polio elimination efforts is we get that push to the end of the road. You can't over-look the kid at the end of the road, because that kid is your reservoir. I think we should focus on people not diseases but by creating the focus on disease you get that focus on the kid that's hard to reach.[13]

The lady-in-waiting

Resources and knowledge are being passed from team to team but the Gates Foundation seems nervous about immediate calls to mount another single disease eradication effort. Lea Hegg expressed doubts about moving on to the most likely candidate, measles. 'There's a certain group that sees the opportunity to take all of the polio staff and shift them to another eradication initiative like measles,' she said:

> But at the country level there's a lot of fatigue over the pace of what we call SIAs [supplementary immunization activities] for polio and a recognition that when you focus on one disease you're neglecting the system that could raise immunity for all of the diseases. At the global level my sense is that donors aren't willing to support another initiative like polio eradication, with all the resources it would need. They just won't get behind the measles effort.

Single disease campaigns have big benefits. Without their energy and focus, Hegg admitted, it is hard to get donors to support the day-in day-out work of routine immunization: 'It gets down to the core question of responsibility. International donors are willing to put money into eradication efforts because it works outside the health system. It's a rallying cry and it's a way to mobilize resources.' Without a big-name disease to rally round, Hegg warned, the money could simply stop, and donors expect countries to fund routine immunization themselves. At the Gates Foundation, Hegg said, the focus is on making sure that investment doesn't dry up or move on. 'But then you have to create that vision at the global level. It's tough.'[14]

Hegg's colleague, Matt Hanson, said the foundation's policy is that it is not *anti*-measles eradication, it is just not yet *pro*-eradication:

> To be crystal clear, our organization is not anti-eradication. We're not against it for measles. But at the present time we think it's premature to go full speed ahead with the measles eradication goal for a variety of reasons. With the polio eradication effort taking up so much energy and resources, the political interest is not there. And neither are the dollars to back it.

For years, measles has been the 'lady-in-waiting,' Hanson said, even though, 'it should be the number one priority of any EPI programme manager, because it's the one that could kill the most of your kids if you don't'. Like Hegg, Hanson admitted the close ties and continuing efforts of the polio programme have sometimes exhausted public health efforts:

> Sometimes measles is done a disservice because it's linked with polio. Some people have a bad taste in their mouth from polio, with the amount that it's cost and the time that it's taken. That has unfairly stained the measles work. Yet when you look at the progress that's been made against measles it's staggering.

Three million children are prevented from dying from vaccine-preventable diseases every year, Hanson said, and half of that figure is due to measles vaccination. Many donors and people in government have either had the disease or know people who have had it, which means that it is not a difficult sell, Hanson said: 'It's just a difficult sell at this time.' Not only is the political climate not conducive to developing a polio-scale measles programme, Hanson believes, but there are also technical issues, such as the weakness of some routine immunization programmes:

> This is where disagreements come up between us and the CDC or WHO. For us, routine immunization coverage, especially in the countries that have the most measles right now, is not at the levels it needs to be for adding on an eradication agenda. If you look at the DRC, for example and the routine immunization coverage they have, could you add on an eradication agenda there when you need well over ninety per cent population immunity from measles to stop it? We don't think that that is reasonable.

Given sufficient progress, Hanson said, a more serious discussion about measles eradication might be on the table by 2020 but 'when we say that to partners they don't like it, because they feel like all they are lacking is money. We disagree with that. Some within the community would say "give us a half a billion dollars and all your wishes will come true" and we just don't agree.'[15]

Despite the wariness at the Gates Foundation, the GPEI polio legacy planning paper boldly set out the case in favour of measles eradication as the next target. The paper proudly noted:

> The initiative has assembled an unprecedented and committed global partnership led by Rotary International, the WHO, UNICEF, the CDC and the Bill & Melinda Gates Foundation, which has collectively and relentlessly worked together to overcome the many challenges the initiative has faced and whose vanguard is the twenty million frontline vaccinators. This largest-ever global health partnership is in an ideal strategic position to move forward on other global health challenges . . . such as the effort to wipe the measles virus off the face of the Earth.[16]

Statements like this reveal that the prospect of eradication is as potent now as ever, even with a hint of madness, sweeping aside dissent and pushing ahead.

CONCLUSION

If we had a magic pill that got rid of all infectious diseases, would we really use it?

Nim Arinaminpathy, Epidemiologist[1]

On the day the eradication of smallpox was announced D. A. Henderson, the man who had driven the campaign on against the odds, admitted he felt a sense of loss. Health ministers from around the world gathered at the World Health Assembly in Geneva to receive the final report from Dr Frank Fenner, who oversaw the smallpox eradication certification process. 'It was a moving ceremony and a memorable moment,' Henderson remembered, but nonetheless sad: 'Never again could I expect to share with so many colleagues the day-to-day excitement, stress, anxiety, elation and satisfaction in achievement that had been our common experience for more than a decade.'[2]

Some day soon, we will know the name of the last child to suffer from polio. A recent study showed that ninety-eight per cent of parents in Pakistan now accept polio vaccines for their children (a higher rate than in some of the US and Europe) while between 2013 and 2015 the number of children who could not be reached by vaccinators fell from half a million to 35,000.

For that final child and their family, who will probably come from the Khyber area of Pakistan, the illness will be a tragedy but it will be a moment of victory for the millions of health workers, in countries around the world, who continued to believe eradication was possible and worthwhile and who often had to be willing to risk everything to deliver a vaccine for a disease most of the world had forgotten about.

We live amid a dense cloud of microbes and viruses; many help us but some make us ill, and in doing so weed out the weakest among us, in a Darwinian selection. What is the value of tackling one disease among so many and pursuing its eradication with fanatical fervour? Eradication campaigns involve huge expense, massive manpower and decades of hard work. For some, they mean using resources they believe could have been better spent else-where, substantial intrusion into their lives and considerable pressure to comply with lofty global goals that seem to have little relevance to their day-to-day existence. And all of them, with the exception of smallpox, have ended in failure.

Are eradication campaigns a form of madness, pursued by healthcare imperialists such as Fred Soper and crazy scientists such as Hideyo Noguchi, who died of yellow fever, the disease he was desperate to conquer? Are they vanity projects for billionaire philanthropists and their foundations, keen to stamp their name next to a concrete achievement? Or are they an undeniable, irre-versible 'win', so rare and elusive in a world of compromise and qualification?

We live in a world that is driven by sharp debates about individ-ual rights, where there is less respect for concepts such as 'herd immunity', and people are careless about other threats to long-term human survival, such as climate change. In such times, can global institutions summon the political support and funding to embark on eradication campaigns whose core is the notion that every human being is equal, regardless of race, nationality, sex, status or financial resources?

As the polio programme races to track down those final few cases, 2016 is becoming the year that brought such arguments to a head; the year of Brexit and Donald Trump versus Hillary Clinton and European unity. A year that illuminated the clash of ideologies between an open world and a closed one where concrete walls attempt to hold frightening threats at bay. In this light, the work of the polio vaccinators belongs as much to the future as to the past. No wall can hold back deadly viruses; that needs the open collaboration of people across the world.

Delivering the Nelson Mandela Annual Lecture at the University of Pretoria in July 2016, Bill Gates outlined the potential dynamism young Africans held for their continent, as well as the health, nutrition, educational and economic challenges holding them back. Without underestimating the scale of those challenges, Gates believed the dynamism could be unleashed, and the near-but-yet-so-far prospect of a polio-free Nigeria pointed to what was possible. Thinking back to one of his earliest conversations with Nelson Mandela, Gates said that his hope for the future resided in the concept that: 'There is a universal appeal to the conviction that youth deserves a chance.'[3]

For millions around the world the uniquely successful campaign against smallpox, as well as the sometimes misguided and often doomed campaigns against diseases such as malaria, yellow fever and polio have given young people the biggest chance of all: the chance to live. That surely remains the dream of disease eradicators everywhere.

ACKNOWLEDGEMENTS

I would like to thank all the experts and health workers who agreed to be interviewed for this book, as well as those who told me their own stories about suffering from diseases such as polio. In particular I would like to thank Rachel Lonsdale and Amber Zeddies at the Gates Foundation, Leah Sandals at Global Health Strategies and Sona Bari at the WHO for spending so much time talking through their programmes with me and for arranging so many interviews.

I would also like to thank my mother for giving me quite heroic support in completing this book, my agent Gaia Banks at Sheil Land and Sam Carter at Oneworld for pulling the manuscript into shape and Amna Khwaja for having input at every stage.

NOTES

This book uses a timeframe of 2014–2016.

1 The Hippies Who Beat Smallpox

1   Author interview with Tim Evans, director for Health, Nutrition and Population at the World Bank.
2   Author interview with Larry Brilliant.
3   Article about Larry Brilliant, 30 September 2000, *Fast Company* by Harriet Rubin.
4   *Angel of Death: The Story of Smallpox* by Gareth Williams, Palgrave Macmillan 2011, p. 12.
5   Edward Jenner quoted in *Angel of Death: The Story of Smallpox*, p. 197.
6   Author interview with Gareth Williams.
7   Thomas Jefferson letter to Edward Jenner quoted in *Angel of Death: The Story of Smallpox*, p. 230.
8   Author interview with Bill Foege.
9   Bill Foege article, *Nature* magazine, 2001.
10 Author interview with Chris Burns-Cox.
11 Ibid.
12 Ibid.
13 Author interview with Larry Brilliant.
14 Author interview with Bill Foege.
15 Author interview with Walter Orenstein.
16 Author interview with Larry Brilliant.

2 The Crippler

1   Author interview with Larry Brilliant.

2   Author interview with Christine Wright.

3   Author interview with Larry Brilliant.

4   Ibid.

5   Author interview with Selamawit Satato.

6   Author interview with doctor (anonymous).

7   Walter Scott quoted in *Polio: An American Story* by David M. Oshinsky, Oxford University Press, 2005, p. 10.

8   *Jonas Salk: A Life* by Charlotte DeCroes Jacobs, Oxford University Press, 2015, prologue.

9   *Time* magazine quoted in 'The President and the Wheelchair' by Christopher Clausen in the *Wilson Quarterly*, Summer 2005.

10 *Polio: An American Story*, p. 54.

11 Ibid.

12 NFIP information film.

13 Hilary Koprowski speaking at the launch of his vaccine, March 1951.

14 Hilary Koprowski obituary, *New York Times*, 20 April 2013.

15 Hilary Koprowski speaking at the meeting on immunization in poliomyelitis held in Hershey, Pennsylvania, 15–17 March 1951.

16 David M. Oshinsky quoted in article about Koprowski, 20 April 2013, *New York Times* by Margalit Fox .

17 Ibid.

18 *Jonas Salk: A Life*, p. 21.

19 Salk interview, 14 November 1990, MOD AV collection, quoted in *Jonas Salk: A Life*, p. 14.

20 *Jonas Salk: A Life*, p. 37.

21 Ibid, p. 92.

22 Ibid, p. 108.

23 Jonas Salk interview with Richard Carter.

24 Richard Mulvaney quoted in 'Polio Vaccine's Proud Pioneers', *The Fairfax Connection*, 28 April 2004.

25 Gail Adams Batt quoted in 'Polio Vaccine's Proud Pioneers' *The Fairfax Connection*, 28 April 2004.

26 *Jonas Salk: A Life*, p. 171.

27 *Paralysed with Fear: the Story of Polio* by Gareth Williams, Palgrave Macmillan,

2013, p. 236.

28 Ibid, p. 237.

29 *Jonas Salk: A Life*, p. 212.

30 Ibid.

3 Polio's Last Stand

1   Author interview with Dr John Sevrer.

2   This and all future quotes author interview with John Sevrer.

3   'Rotary and the Gift of a Polio Free world' by Sarah Gibbard Cook, Rotary Club Report, 2013, p. 68

4   Ibid.

5   Author interview with John Sevrer.

6   Author interview with Chris Maher, May 2014.

7   Author interview with Hamid Jafari, May 2014.

8   Ibid.

9   Author interview with Naveen Thacker, May 2014.

10 Author interview with Chris Maher.

11 Ibid.

12 Author interview with Naveen Thacker.

13 Ibid.

14 Author interview with Chris Maher.

15 Author interview with Naveen Thacker.

16 Author interview with Chris Maher.

17 Author interview with Sumita Tharpar.

18 Author interview, Indian government, May 2014.

19 Author interview with Chris Maher.

20 Author interview with Hamid Jafari, May 2014.

21 Author interview with Peter Crowley, UNICEF, May 2014. All future Crowley quotes attributed to this.

22 Author interview with Sir Liam Donaldson, 2014.

23 Statement by Lisa Monaco, President Obama's senior counterterrorism and homeland security adviser, May 2014.

24 Author interview with Rebecca Martin, CDC, June 2014.

25 Author interview with Peter Crowley.

26 Author interview with Rebecca Martin.

27 Author interview with Hamid Jafari.

28 Author interview with Aziz Memon, 2014.

29 Author interview with Peter Crowley.

30 Author interview with Aziz Memon, 2014.

31 Author interview with Tim Petersen, Gates Foundation, October 2015.

32 Author interview with Tim Petersen.

33 Author interview Hamid Jafari, WHO, August 2015.

34 BBC report, 30 January 2016.

35 Author interview Hamid Jafari, August 2015.

36 Author interview Tim Petersen.

4: Bill Gates and the Final One Per Cent

1 Bill Gates, Nelson Mandela Lecture, University of Pretoria, 17 July 2016.

2 Author interview with Michael Galway, October 2015.

3 Author interview with Greg Armstrong, CDC, June 2014.

4 Author interview with Mulugeta Debesey, May 2014.

5 Author interview with Brigitte Toure, May 2014.

6 Author interview with Craig Allen, CDC, June 2014.

7 Author interview with Michael Galway, October 2015.

8 Author interview with Dr Yagoub Al Masrou, the Islamic Advisory Group (IAG), summer 2014.

9 Unicef report, 2014.

10 Author interview with Michael Galway.

11 Author interview with Greg Armstrong, CDC, June 2014.

12 Author interview with Michael Galway.

13 Alex Kornblum, 'Can a Costly Campaign to Eradicate Polio From Nigeria Possibly Succeed?' The Nation, 25 November 2014.

14 Author interview Craig Allen, CDC, June 2014.

15 Author interview with Oyewale Tomori, summer 2014.

16 Author Interview with Michael Galway.

17 Alex Kornblum, *The Nation*, 25 November 2014.

18 Author interview with Oyewale Tomori.

19 Author interview with Michael Galway.

20 Ibid.

21 Ibid.

22 Author interview with Vio Mitchell, October 2015.

23 Alex Kornblum, *The Nation*, 25 November 2014.

24   Jeff Goodell, *Rolling Stone*, March 2014.

25   Author interview with Apoorva Mallya, October 2015.

26   Author interview with Sir Liam Donaldson, May 2014.

27   Author interview with Jay Wenger, May 2014.

28   Author interview with Chris Maher.

29   Author interview with Carol Pandak, Rotary, May 2014.

30   Author interview with Vio Mitchell, October 2015.

31   Author interview with Apoorva Mallya, October 2015.

32   Author interview with Walter Orenstein, June 2014.

33   Author interview with Sir Liam Donaldson.

34   Author interview with Walter Orenstein, June 2014.

35   Author interview with John Modlin, October 2015.

36   Author interview with Jay Wenger.

37   Author interview with Walter Orenstein, June 2014.

38   Author interview John Vertefeuille, current incident manager for polio at the CDC, October 2015.

39   Author interview with Hamid Jafari, London, May 2014.

5: The Prophet

1   'The Mosquito Killer', article about Fred Soper, by Malcolm Gladwell, *New Yorker*, 21 July 2001.

2   *New York Times*, 31 January 2011.

3   *The End of Plagues: The Global Battle Against Infectious Disease* by John Rhodes, Macmillan 2013, p. 83.

4   Ibid, p. 89.

5   *Eradication: Ridding the World of Diseases Forever?* by Nancy Leys Stepan, Reaktion Books, 2011, p. 24.

6   Ibid, p. 35.

7   Samuel Breck's account of a yellow fever outbreak in Philadelphia, 1793, appears in Hart, Albert Bushnell, *American History Told by Contemporaries*, vol. 3, 1929.

8   *Mosquito: The Story of Man's Deadliest Foe* by Andrew Spielman and Michael D'Antonio, Faber and Faber 2001, p. 13.

9   *Eradication*, Nancy Leys Stepan, p. 57.

10   Ibid.

11   Ibid, p. 58.

12 Ibid, p. 49.

13 Quoted in *Eradication*, Nancy Leys Stepan, p. 63.

14 'The Mosquito Killer', article about Fred Soper by Malcolm Gladwell, *New Yorker*, 21 July 2001

15 'Yellow Fever in the Anglo-Egyptian Sudan' by Heather Bell. This article originally appeared in the 1995 *Rockefeller Archive Center Newsletter*.

16 Paul Russell quoted in Gordon Patterson, *The Mosquito Crusades*, Rutgers University Press, 2009.

17 'The Mosquito Killer' by Malcolm Gladwell, *New Yorker*, 21 July 2001.

18 *Mosquito: The Story of Man's Deadliest Foe*, p. 13.

19 *Eradication*, Nancy Leys Stepan, p. 102.

20 'The Mosquito Killer' by Malcolm Gladwell, *New Yorker*, 21 July 2001.

6: The Mosquito House

1 Author interview with Dr Peter Agre, January 2016.

2 Ibid.

3 Ibid.

4 Author interview with Hugh Sturrock, January 2016.

5 WHO statement by Dr Margaret Chan.

6 *Mosquito* Andrew Spielman, p. 14.

7 Ibid, p. 88.

8 *Eradication*, Nancy Leys Stepan, p. 145.

9 *Eradication*, Nancy Leys Stepan, p. 158.

10 *Silent Spring* by Rachel Carson, Houghton Miffin, 1962.

11 *The Fever* by Sonia Shah, Picador, 2010, p. 210.

12 Author interview with Sir Richard Feachem, January 2016.

13 Bhatt, S., et al. (2015) 'The effect of malaria control on *Plasmodium falciparum* in Africa between 2000 and 2015', *Nature* 526 (7572) 207-211. Available from www.nature.com/nature/journal/v526/n7572/full/nature15535.html

14 *The Economist*, November 2015.

15 Ibid.

16 Author interview with Sir Richard Feachem, January 2016.

17 *The Fever* by Sonia Shah, p. 113.

18 Author interview with Sir Richard Feachem, January 2016.

19 Ibid.

20 Ibid.

21   Author interview David Schellenberg, London School Hygiene and Tropical Medicine January 2016.

22   Ibid.

23   Author interview with Sir Richard Feachem, January 2016.

24   Author interview with Janet Hemingway, Liverpool School Tropical Medicine, January 2016.

25   Author interview with Janet Hemingway

26   Author interview with Sir Richard Feachem, January 2016.

27   Author interview with Janet Hemingway, January 2016.

28   Author interview with Sir Richard Feachem, January 2016.

29   Author interview with Hugh Sturrock, January 2016.

30   Ibid.

31   Author interview Ivan Gayton, MSF, January 2016.

32   Author interview with Sir Richard Feachem, January 2016.

33   Ibid.

34   *Mosquito* by Andrew Spielman, preface.

7: Liberty or Death – The Anti-vaccination Movement

1   George Bernard Shaw, *The Doctor's Dilemma*, 1906.

2   *On Immunity: An Inoculation*, by Eula Biss, Fitzcarraldo Editions, 18 February 2015, p. 31.

3   Study by Andrew Wakefield and 12 colleagues published in *The Lancet*, 1998.

4   AP reporting 26 June, 2015.

5   Ibid.

6   *The Angel of Death* by Gareth Williams, p. 235.

7   *The Angel of Death* by Gareth Williams, p. 239.

8   International Anti-vaccination League pamphlet.

9   *The Angel of Death*, p. 257.

10   *The Angel of Death*, p. 258.

11   *Paralysed by Fear*, by Gareth Williams, p. 259.

12   Author interview with Elena Conis, February 2016.

13   *Vaccine Nation: America's Changing Relationship with Immunization*, by Elena Conis, University of Chicago Press, 2015, p. 2.

14   Author interview with Elena Conis.

15   *Vaccine Nation*, p. 91.

16   Ibid, p. 106.

17 Ibid, p. 123.

18 *On Immunity*, p. 54.

19 Daniel Lander quoted in *Vaccine Nation*, p. 139.

20 Author interview with Elena Conis.

21 William Saletan, 'Sexually Transmitted Injection', *Slate*, 15 October 2009.

22 Author interview with Elena Conis.

23 WHO *Bulletin*, 2 July 1986.

24 Ibid.

25 Author interview with Steven Salzberg, February 2016.

26 Author interview with Elena Conis.

27 'Applying Behavioral Economics to the Thorny Issue of Vaccine Acceptance', Rebecka Rosenquist, *Wharton Magazine*, University of Pennsylvania, 22 August 2016.

8: Ebola: An Avoidable Crisis

1 *Ebola: The Natural and Human History of a Deadly Virus* by David Quammen, W. W. Norton and Company, updated 2014.

2 Author interview with Dr Failsal Shuaib.

3 Author interview with Hamid Jafari, August 2015.

4 Ibid.

5 Author interview with Peter Agre, January 2016.

6 Author interview with Angel Hwang, October 2015.

7 *No Time To Lose: A Life in Pursuit of Deadly Viruses* by Peter Piot, W. W. Norton and Company, 2012, p. 3.

8 Ibid, p. 42.

9 Ibid, p. 57.

10 *Ebola*, David Quammen, p. 24.

11 Ibid, p. 25.

12 Ibid, p. 48.

13 Ibid, p. 39.

14 *New England Journal of Medicine*, October 2014, Peter Piot and Jeremy Farrar.

15 'How Ebola Roared Back' by Kevin Sack, Sheri Fink, Pam Belluck and Adam Nosster, *New York Times*, 30 December 2014.

16 WHO figures, 13 January 2016.

17 *Ebola*, David Quammen, p. 110.

18 'How Ebola Roared Back'.

19   Ibid.

20   *Guardian*, 28 April 2016. 'Ebola Leaves a Painful Legacy in Sierra Leone', Olivia Acland.

9: The Legacy

1   'The Global Polio Eradication Initiative: Progress, Lessons Learned and Polio Legacy Transition Planning' by Stephen Cochi, Lea Hegg, Anjali Kaur, Carol Pandak and Hamid Jafari, February 2016.

2   Author interview with Ashok Maloo, August 2015.

3   Radio interview, PRI *The World*, 21 August 2015.

4   Ibid.

5   Author interview with Ashok Maloo, August 2015.

6   'The Global Polio Eradication Initiative: Progress, Lessons Learned and Polio Legacy Transition Planning'.

7   Author interview with Lea Hegg, October 2015.

8   Author interview with Ivan Gayton, MSF.

9   'Current Trends of Immunization in Nigeria', *Journal of Tropical Medicine and Health*, June 2014. Endurance A. Ophori, Musua Y. Tula, Azuka V. Azih, Precious E. Ikpo.

10   Author interview with Angela Hwang, October 2015.

11   Ibid.

12   'The Global Polio Eradication Initiative: Progress, Lessons Learned and Polio Legacy Transition Planning'.

13   Author interview with Angela Hwang.

14   Author interview with Lea Hegg.

15   Author interview with Matt Hanson, October 2015.

16   'The Global Polio Eradication Initiative: Progress, Lessons Learned and Polio Legacy Transition Planning'.

Conclusion

1   Nim Arinaminpathy, Epidemiologist. Livescience, 'What If We Eradicated All Infectious Disease?', Natalie Wolchover, 8 June 2012.

2   *Smallpox: The Death of a Disease : The Inside Story of Eradicating a Worldwide Killer*, D. A. Henderson, Prometheus Books, 2009, p. 249.

3   Bill Gates, Nelson Mandela Memorial Lecture, University of Pretoria, 17 July 2016.

INDEX

References to images are in italics.

Index